AN ARTIFICIAL INTELLIGENCE
APPROACH TO VLSI DESIGN

THE KLUWER INTERNATIONAL SERIES
IN ENGINEERING AND COMPUTER SCIENCE

KNOWLEDGE REPRESENTATION, LEARNING AND EXPERT SYSTEMS

Consulting Editor
Tom M. Mitchell

AN ARTIFICIAL INTELLIGENCE APPROACH TO VLSI DESIGN

THADDEUS J. KOWALSKI
Bell Telephone Laboratories, Inc.
Murray Hill, New Jersey

KLUWER ACADEMIC PUBLISHERS
Boston/Dordrecht/Lancaster

Distributors for North America:
Kluwer Academic Publishers
190 Old Derby Street
Hingham, MA 02043, USA

Distributors outside North America:
Kluwer Academic Publishers Group
Distribution Centre
P.O. Box 322
3300 AH Dordrecht
THE NETHERLANDS

Library of Congress Cataloging in Publication Data

Kowalski, Thaddeus J.
  An artificial intelligence approach to VLSI design.

  (The Kluwer international series in engineering and
computer science; SECS 4)
  Bibliography: p.
  Includes index.
  1. Integrated circuits — Very large scale integration.
2. Artificial intelligence. 3. Expert systems.
4. Digital electronics. I. Title. II. Series.
TK7874.K675 1985   621.3819'5835   85-7581
ISBN 0-89838-169-X

Copyright © 1985 by Bell Telephone Laboratories, Inc.

Third printing, 1986

All rights reserved. No part of this publication may be reproduced, stored in a retrieval system, or transmitted in any form or by any means, mechanical, photocopying, recording, or otherwise, without written permission of the publisher, Kluwer Academic Publishers, 190 Old Derby Street, Hingham, Massachusetts 02043.

Printed in the United States of America

To William Shelley, my fourth grade teacher . . .
He made learning fun.

CONTENTS

LIST OF FIGURES	ix
LIST OF TABLES	x
PREFACE	xi
ACKNOWLEDGEMENT	xiii

1.	INTRODUCTION	1
	1.1 MOTIVATION	2
	1.2 CMU/DA LEVELS	3
	1.3 PROBLEM STATEMENT	4
	1.4 APPROACH	5
	1.5 RELATED RESEARCH	5
	1.6 OVERVIEW	7
2.	KNOWLEDGE-BASED EXPERT SYSTEMS	9
	2.1 GENERAL COMPONENTS OF A KNOWLEDGE-BASED EXPERT SYSTEM	10
	2.2 GENERAL FEATURES OF A KNOWLEDGE-BASED EXPERT SYSTEM	12
	2.3 EXAMPLE KNOWLEDGE-BASED EXPERT SYSTEMS	12
	2.4 KBES SUMMARY	15
3.	DAA DEVELOPMENT	16
	3.1 METHODS TO ACQUIRE EXPERT KNOWLEDGE	17
	3.2 THE CASE-STUDY INTERVIEWS	18
	3.3 PROTOTYPE SYSTEM	23
	3.4 KNOWLEDGE ACQUISITION INTERVIEWS	26
	3.5 ANALYSIS OF KNOWLEDGE	31
	3.6 DEVELOPMENT SUMMARY	39
4.	DAA REPRESENTATIONS	41
	4.1 ALGORITHMIC REPRESENTATION	42
	4.2 TECHNOLOGY DATABASE AND CONSTRAINTS	47
	4.3 TECHNOLOGY-INDEPENDENT HARDWARE NETWORK	51
	4.4 BOOKKEEPING INFORMATION	55
	4.5 REPRESENTATION SUMMARY	56
5.	DAA KNOWLEDGE	58
	5.1 SERVICE FUNCTIONS	61
	5.2 GLOBAL ALLOCATION	63
	5.3 VALUE TRACE ALLOCATION	66
	5.4 SCS ALLOCATION	72
	5.5 ESTIMATORS	77
	5.6 GLOBAL IMPROVEMENTS	82
	5.7 KNOWLEDGE SUMMARY	86

6. THE IBM SYSTEM/370 EXPERIMENT	**89**
6.1 THE D370 DESIGN	90
6.2 THE μ370 DESIGN	96
6.3 THE D370 AND μ370 DESIGN COMPARISON	99
6.4 SUMMARY OF THE SYSTEM/370 EXPERIMENT	102
7. CONCLUSION	**103**
REFERENCES	**106**
APPENDIX A: WORKING-MEMORY VT	**115**
APPENDIX B: WORKING-MEMORY USER PARAMETERS	**118**
APPENDIX C: WORKING-MEMORY SCS	**123**
APPENDIX D: THE SYSTEM/370 CRITIQUE	**126**
INDEX	**208**

LIST OF FIGURES

Figure 1.	MCS6502 — 8 BIT MUX DATA PATHS	29
Figure 2.	MCS6502 — 8 BIT BUS DATA PATHS	30
Figure 3.	MODULE-TO-MUX-AND-BUS-WITH-BUS-TO-MUX	32
Figure 4.	ASSIGN-INC	33
Figure 5.	BIND-ARITHMETIC-OPERATORS-SAME-SIZE	34
Figure 6.	ALLOCATION	34
Figure 7.	NEXT	35
Figure 8.	DEFAULT-PLUS	35
Figure 9.	DEFAULT-FOLD-ARITHMETIC	36
Figure 10.	CLEAN-OUTNODES	36
Figure 11.	DECLARED-VARIABLE-ALLOCATION-DONE	36
Figure 12.	TRIM-ADD-SUB-MUL-OUTPUT-ZERO-NIL-TRIM	37
Figure 13.	TRIM-OUTPUT	37
Figure 14.	NEXT-LIST-1	38
Figure 15.	END-LIST-1	38
Figure 16.	FOLD-REGISTER-PROX	39
Figure 17.	NOT-STABLE-REGISTERS	39
Figure 18.	SAMPLE ISPS DESCRIPTION	42
Figure 19.	SAMPLE VT	43
Figure 20.	SAMPLE STRUCTURAL SPECIFICATION	52
Figure 21.	SAMPLE CONTROL SPECIFICATION	53
Figure 22.	DAA SUBTASKS	59
Figure 23.	MAKE LINK	63
Figure 24.	GLOBAL ALLOCATION	64
Figure 25.	DESIGN AFTER GLOBAL ALLOCATION	65
Figure 26.	FIND-VTBODIES-FOR-REGISTERS	66
Figure 27.	VT ALLOCATION	67
Figure 28.	DESIGN AFTER VALUE TRACE ALLOCATION	68
Figure 29.	FIND-OUTNODES-FOR-TEMPORARIES	70
Figure 30.	SCS ALLOCATION	73
Figure 31.	DESIGN AFTER SCS ALLOCATION	74
Figure 32.	FOLD-NO-OUTPUT-REGISTER	75
Figure 33.	GLOBAL IMPROVEMENTS	83
Figure 34.	DESIGN AFTER GLOBAL IMPROVEMENTS	84
Figure 35.	CONVERT-MUX-INPUTS-TO-BUS	85
Figure 36.	AN EXAMPLE BUS ALLOCATION	86
Figure 37.	THE D370 DESIGN — PART 1	91
Figure 38.	THE D370 DESIGN — PART 2	92
Figure 39.	THE μ370 DESIGN	97

LIST OF TABLES

Table	1.	A DECADE OF SYSTEMS	13
Table	2.	A DECADE OF TOOLS	14
Table	3.	SUMMARY OF INTERVIEWS — PART 1	19
Table	4.	SUMMARY OF INTERVIEWS — PART 2	20
Table	5.	MCS6502 — THREE DESIGNS	28
Table	6.	RULES BY KNOWLEDGE TYPE	31
Table	7.	VTBODY	45
Table	8.	OUTNODE AND OPERATOR	46
Table	9.	BRANCHES, LISTS AND TREES	47
Table	10.	USER PARAMETERS	48
Table	11.	DEFAULT DB-OPERATOR — PART 1	49
Table	12.	DEFAULT DB-OPERATOR — PART 2	50
Table	13.	DEFAULT DB-OPERATOR — PART 3	51
Table	14.	DEFAULT FOLD AND DELAY	51
Table	15.	HARDWARE	54
Table	16.	MISCELLANEOUS	56
Table	17.	RULES BY FUNCTION AND KNOWLEDGE TYPE	60
Table	18.	MEMORY ARRAYS IN THE D370 DESIGN	94
Table	19.	REGISTERS IN THE D370 DESIGN	95
Table	20.	IBM SYSTEM/370 — DESIGN DIFFERENCES	100

PREFACE

VLSI design synthesis is a method for designing hardware that starts with an algorithmic description and uses interactive computer programs to create a finished design. This structured approach can decrease the time it takes to design a chip, automatically provide multi-level documentation for the finished design, and create reliable and testable designs. VLSI design synthesis is a difficult problem because the huge number of facts and implicit dynamic constraints do not lend themselves to a recipe-like solution. However, the Knowledge Based Expert System, KBES, approach provides a framework for organizing solutions to problems that are currently solved by experts using large amounts of domain-specific knowledge. Therefore, the goals of my research are to extract and codify the knowledge of expert designers for a better understanding of VLSI design-synthesis, to implement a KBES capable of creating efficient, testable and usable designs, and to determine the usefulness of the KBES technique for implementing design automation tools.

A series of acquisition interviews and an initial prototype system have been used to bootstrap a system that generates a technology-independent list of operators, registers, data paths and control signals from an algorithmic description. The Design Automation Assistant, DAA, has been used to design an IBM System 370 and was favorably evaluated by an IBM System/370 designer.

The advantages of using the KBES approach are the ease of validating the knowledge gathered from interviews with experts, the ease of incrementally adding to the knowledge base, the ability to query that knowledge during the design task, and the replacement of extensive backtracking by domain-specific knowledge techniques. The disadvantage is the difficulty of extracting knowledge from the experts.

This thesis adds to knowledge in the digital design synthesis domain by compiling and testing the set of rules used by expert designers, and to knowledge in the expert system domain by providing another system for researchers to examine. This knowledge will also aid in the teaching of design by making explicit knowledge that is now passed on primarily through apprenticeship.

ACKNOWLEDGEMENT

First and foremost, my profound thanks go to my parents, Henry and Ann Kowalski. None of this could have been done without their indulgence, support, enthusiasm, and love. Thank you.

My sincere thanks goes to my adviser, Donald Thomas for his patience and support through many phone calls and meetings. I would like to thank Stephen Director, Allen Newell, John McDermott, and Daniel Siewiorek for their encouragement, support, and critical review of my research. Also, the thoughtful criticisms and suggestions of friends and colleagues have added greatly to this thesis and to my pleasure in writing it. In particular, David Ditzel, Mitch Marcus, Thomas Mitchell, Lisa Masakowski, and Myron Wish all read multiple versions with care. Marcia Derr, Charles Forgy, Karen Kukich, Walter MacWilliams, Michael McFarland, Sharon Murrel, Peg Schafer, Sherri Shulman, and Sandra Pruzansky have made helpful comments at various stages of the research. The many members of the CMU/DA community have provided many ideas and probing questions, especially David Gatenby, David Geiger, Charles Hitchcock, John Lertola, Matt Mathis, John Nestor, Jayanth Rajan, and Robert Walker. Invaluable assistance with the mechanics of my thesis was provided by the UNIX operating system, the TROFF text processing program, the writers workbench, and the OPS5 knowledge-based expert-system writing tool.

I would like to thank Ken Chong, Claud Davis, David Ditzel, Michael Maul, Gil Mowery, Allen Ross, Clayton Schneider, Glen Williams, and Andrew Wilson for donating time to critique the various designs. Lastly, I would like to thank the INTEL and IBM corporations for providing designers to critique the designs, with special thanks to AT&T Bell Laboratories for computer resources and designer's critiques. This work was supported in part by an IBM fellowship and the National Science Foundation.

If I Only Had A Brain

I could while away the hours
conferrin' with the flow'rs
consultin' with the rain
And my head, I'd be scratchin'
while my thoughts were busy hatchin'
If I Only Had A Brain.

I'd unravel ev'ry riddle
for any individle
in trouble or in pain
With the thoughts I'd be thinkin'
I could be another Lincoln,
If I Only Had A Brain.

Oh, I could tell you why the ocean's near the shore,
I could think of things I never thunk before
and then I'd sit and think some more.

"IF I ONLY HAD A BRAIN" Lyric by E. Y. Harburg, Music by Harold Arlen © 1938, Renewed 1966 METRO-GOLDWYN-MAYER INC. © 1939, Renewed 1967 LEO FEIST, INC. Rights Assigned to CBS CATALOGUE PARTNERSHIP. All Rights Controlled & Administered by CBS FEIST CATALOG INC. All Rights Reserved. International Copyright Secured. Used by Permission.

Chapter 1

INTRODUCTION

Recent advances in integrated circuit fabrication technology have allowed larger and more complex designs to form complete systems[1] on single VLSI chips. These chips use one-micron to five-micron features to achieve complexities equivalent to 100,000 to 250,000 transistors. This level of design complexity has created a combinatorial explosion of details — a major limitation in realizing cost-effective, low-volume, special-purpose VLSI systems. To overcome this limitation, design tools and methodologies capable of automating more of the digital synthesis process must be built.

We have been developing just such synthesis tools[2,3] in the Carnegie-Mellon University design automation, CMU/DA, community. These tools help the designer develop the algorithmic description of the system and interactively add the details required to produce a finished design. This structured approach can decrease the time it takes to design a chip, automatically provide multi-level documentation for the finished design, and create reliable and testable designs.

This thesis focuses on the synthesis, or allocation, of the implementation-design space as it advances from an algorithmic description of a VLSI system to a list of technology-independent registers, operators, data paths and control signals. Our approach is aimed at aiding the designer by producing data paths and control sequences that implement the algorithmic system description within supplied constraints. Thus, the designer can consider many alternatives before deciding on a final design.

This task has inspired a variety of approaches, ranging from the most simplistic backtracking methods through the most complicated constraint propagation methods.[4,5,6,7,8] Owing to the complexity of design synthesis, simplistic backtracking schemes consume large amounts of CPU time, and the constraint propagation method is too cumbersome for large designs. Because of the combinatorial explosion of details and implicit dynamic constraints involved in choosing an implementation, this problem does not lend itself to these algorithmic solutions. An alternate approach to design synthesis uses a large amount of design knowledge to eliminate backtracking; whenever possible, the focus is on specific design details and constraints. Artificial intelligence researchers have called systems developed under this heuristic approach knowledge-based expert systems, KBESs.[9] This chapter motivates the research, provides background on the CMU/DA system, lists the related research in the area, states the problem, and provides a road map of the rest of the thesis.

1.1 Motivation

If Moore's law[10] continues to hold, within this decade the size of the smallest VLSI feature will be reduced ten times. This increasing potential to fabricate more complex systems will cause at *least* a linear increase in information that must be managed. From information management studies of large software projects,[11] it can be shown that the time to design and implement future VLSI systems will be considerably more than ten times that required at present, which is prohibitive for low-volume special-purpose VLSI systems. It can be shown further that as the cost of designing special-purpose VLSI systems decreases, the demand for expert designers will increase. Thus, design methods should be developed to deal effectively with the magnitude and complexity of VLSI design. Such methods would increase the potential productivity of designers, while also making it possible for more people to design systems that are both testable and reliable. These are areas even good designers often forget, but can mean the difference between working and marginal designs.

Many people are working on problems related to computer-aided design, or CAD, of VLSI systems. For example, if we look in the proceedings of the Twentieth Design Automation Conference, we see topics of hardware description languages, testing, simulation, layout and placement, PLA minimization, and synthesis, to name a few. Although synthesis was considered to be strictly in the realm of the creative designer, automatic and computer-aided synthesis programs are now being developed for many levels of VLSI system design. In general, the quality of the designs produced by automatic synthesis programs is not adequate for complete automation of the design process for production use. However, these programs are beginning to find use as an interactive aid during the design process.[2, 12, 13, 14] Toward this end the development of synthesis tools to aid in the creative design process has become an important area of research.

1.2 CMU/DA Levels

At CMU we look at synthesis as the creation of a detailed representation from an abstract representation. The VLSI design synthesis task can be decomposed into several subtasks, each providing an increasing level of detail from the abstract representation. This section describes the four levels of increasing detail used in the CMU/DA system.[2]

1.2.1 The algorithmic level. The first level is an algorithmic description of the design. At this level of detail the high-level intent of hardware can be understood and simulated[15] regardless of the target design style (for example pipeline, multiplexer, microprocessor, or bus) or technology. This level is represented in the instruction set processor language, ISPS,[16] and a value trace data and control flow description language, VT.[17, 18] ISPS is a programming language similar to ALGOL, while VT is an extraction of the data flow and control flow information present in the designer's ISPS description. VT is easy for computer programs to manipulate and is felt to be less sensitive to different styles of writing the same algorithmic description.

1.2.2 The technology-independent-hardware-network level. The second level is a technology-independent-hardware-network description of the design. At this level of detail the functionality and connectivity of the hardware can be understood regardless of the target design technology (TTL, ECL, NMOS, or CMOS). This level is represented in the technology-independent-hardware-network language, SCS,[19] which describes technology-independent registers, operators, data paths, and control signals.† Data-memory allocation and control allocation are defined as the creation of this level from the algorithmic level while applying classical compiler optimizations[21] and design styles.[22]

1.2.3 The technology-dependent-hardware-network level. The third level is a technology-dependent-hardware-network description of the design. At this level of detail the logic and circuits of the hardware can be understood and simulated[23] regardless of physical placement. This level is represented in the technology-dependent-hardware-network language, DIF,[20] which describes feature assignment detail. Feature binding is the creation of this level from the technology-independent level by selection of registers, operators, data paths and control signals that match available feature data-base entries.[24, 25, 26]

1.2.4 The fabrication-dependent-hardware-network level. The fourth level is a fabrication-dependent-hardware-network description of the design. At this level of detail the physical placement of hardware can be understood and simulated.[27] This level is represented in a fabrication-dependent-hardware-network language, which describes feature placement assignment detail. Layout is the creation of this level from the technology-dependent level using geometric information to guide routing and placement of features.

1.3 Problem Statement

Now that we understand how the CMU/DA system divides the task into synthesis levels, let us examine how expert VLSI designers make the transition from the algorithmic description to a hardware implementation. This thesis examines how expert VLSI designers choose a hardware implementation for MOS-microcomputers and whether a KBES can mimic their results. The goals of the research are to extract, codify and test the knowledge of expert designers for a better understanding of VLSI design-synthesis, to implement a KBES capable of creating efficient, testable and usable designs, and to determine the usefulness of the KBES technique for implementing design automation tools. This adds knowledge in the digital design synthesis domain by compiling and testing the set of rules used by expert designers, and to knowledge in the expert system domain by providing another system for researchers to examine. This knowledge will also aid in the teaching of design by making explicit knowledge that is now

---

† SCS is soon to be replaced by DIF.[20]

passed on primarily through apprenticeship.

1.4 Approach

Through a series of detailed structured interviews, the knowledge that expert VLSI designers use to go from the algorithmic level to the technology-independent hardware-network level has been extracted. The codified knowledge has been tested for completeness and correctness by implementing an interactive system, DAA, that adds register, operator, data path and control signal detail to the VT description of hardware to form the SCS description. DAA is implemented as a production system using the OPS5[28] KBES writing system. This KBES tool is based on the premise that humans solve problems by recognizing familiar sub-problems and apply past solutions. This domain is appropriate for a KBES because there are human experts available whose knowledge has been gained through experience and who can teach this knowledge through apprenticeship. Furthermore, the knowledge required to do the task is extensive and requires the type of organization provided by the KBES approach. The advantages of using the KBES approach are the ease of validating the knowledge gathered from interviews with experts, the ease of incrementally adding to the knowledge base, the ability to query that knowledge during the design task, and the replacement of extensive backtracking by domain-specific knowledge techniques. The disadvantage is the difficulty of extracting knowledge from the experts.

The DAA system has been used to design many computers including the MOS Technology Incorporated MCS6502 and the IBM System/370. The design and many redesigns of the MCS6502 provided stimulus for critiquing the implementation design knowledge contained in DAA. A design of each of the small ISPS descriptions maintained at CMU (RK11, HP21MX, F8, I8080, MOD91, 1802, AM2903, H316, VIDEO, TI1200, PDPTTY, SCF3, PDP4, FP, AM2910, AM2901, ELEV, CHANGE, PQ, MINI, MINIS, AM2909, F9407, AM2902, MARK1) showed that the system would produce a functionally correct design for a large number of test cases. Finally, the knowledge in DAA was tested for generality and robustness by designing a much larger processor, with a completely different instruction set, than DAA had ever seen before, and critiquing this design with a designer not used to develop the knowledge base.

1.5 Related Research

The work done thus far in digital design synthesis is reminiscent of the early chess playing programs.[29] It is easy to get a computer to play a legal game of chess, but it is exceeding difficult to get it to play a good game of chess. The CMU/DA project, like the legal game of chess, gets a computer to generate a working piece of hardware from an algorithmic description.[30] Previous attempts at hardware allocators have either produced designs of unacceptable quality, or have been too expensive computationally even for small designs. This section discusses the shortcomings of the current hardware allocators.

At first glance the problem of hardware allocation appears to be a straightforward extraction of registers, operators, and connections from an algorithmic description, but this overlooks the use of multiple design styles and important constraints like cost, size, power and speed. This classical compiler approach was employed by one allocator[6] which used multiplexer-style data flow; it was found to produce designs that were considerably worse than those made by humans.[31]

The MIMOLA Software System, MSS,[4] is an interactive computer-aided design system that takes into consideration hardware usage and limits on the amount of hardware. The MSS allocator starts with a maximally parallel algorithmic description and serializes it until the limits on the amount of hardware can be met. It does this using a single design style and user-provided hardware restrictions. The MSS allocator makes tradeoffs one line at a time within the algorithmic description. Thus, it is limited to local optimizations.

Further research has taken a constraint language approach to the allocation problem,[5] which uses a mixed-integer linear programming solution technique. Though this approach does not have the previous shortcomings, it allows only trivial problems to be solved because its solution method has non-polynomial hard complexity characteristics.[32]

A graph theoretic approach to digital data path synthesis has been developed in a program called Facet.[8] It first develops a maximally parallel dependency representation. Then it allocates the minimum number of storage elements, data operators, and interconnections units for each disjoint cluster, or clique, of algorithmic operators and values in the design. This approach does not consider testability, ease of layout, or the flexibility of mixing alternate design styles.

Another algorithmic approach has been developed in a program called EMUCS.[7] The algorithm aims at an implementation at minimum "cost" for any quantitative parameter such as power or chip area. The algorithm allocates, in a step by step fashion, processors, registers, and multiplexer and bus transfer paths. The synthesis algorithm, as proposed by McFarland,[33] is iterative in nature. It first analyzes the existing intermediate data path to decide what hardware would potentially be the least costly to bind, not according to the lowest cost during this step, but because it might minimize the costs in the next iteration. It then binds that one element, changing the data path as necessary and iterates, reanalyzing and binding, until all elements have been bound and the final data path created. The designs it creates are good locally, but could be improved with a global view including layout, testability and reliability.

Palladio[34] is a circuit design environment for experimenting with methodologies and knowledge-based expert-system design aids. Its framework is based on the premise that circuit designers need an integrated design environment ranging from implementation simulators to layout generators. Their goal is to provide the means for explicitly representing, constructing and testing design tools and languages. This environment integrates both design tools and design specification languages and provides the conceptual framework required by such integration. It is an attempt to explore how to build circuit design environments rather than an automated synthesis program.

1.6 Overview

This chapter has discussed the growing complexity of the VLSI design synthesis task brought on by the decrease in cost and increase in ability to fabricate VLSI systems, and the limited number of expert VLSI designers. It has overviewed the CMU solution to the problem, specifically highlighting the area of implementation synthesis and the various attempts to solve this problem. It discussed an alternate approach to design synthesis, which uses a large amount of design knowledge to eliminate backtracking; whenever possible, the focus is on specific design details and constraints. This approach creates flexible and testable designs that are sensitive to technology, cost, and layout considerations.

Chapter 2, which provides an overview of the knowledge-based expert-system technique, can be skipped by readers familiar with this approach. Chapter 3 discusses the development cycle of DAA, including the interviews, which led to a prototype system, and the further interviews that have developed the DAA system. Chapters 4 and 5 provide insight into how VLSI systems can be designed by discussing the data representations and the task knowledge used by DAA. This knowledge is tested in Chapter 6 with an IBM System/370 design and critiqued by an expert designer at IBM. Finally Chapter 7 summarizes the thesis and discusses possible future research. In summary, this thesis focuses on the synthesis of the implementation-design space and provides insight into how a KBES approach can be used to design cost-effective, low-volume, special-purpose VLSI systems.

Chapter 2

KNOWLEDGE-BASED EXPERT SYSTEMS

During the past decade KBESs have been developed by researchers in artificial intelligence to help solve problems whose structures do not lend themselves to recipe-like solutions. These systems differ from previous efforts in problem solving by effectively coping with the enormous search spaces of alternatives found in real-world problems. The key to success in KBESs is the ability to use domain-rich knowledge to recognize familiar patterns in the current problem state and act appropriately. This tool is based on the premise that humans solve problems by recognizing one of many familiar patterns in the current situation and applying the appropriate actions when this pattern occurs. Their recognition of the pattern is not based on the complete current situation, nor is it recognized with absolute certainty. The presence and absence of patterns can be used to help establish and rule out actions for a given situation.

Researchers have developed many KBESs whose features[35] have matured and grown into usable engineering tools.[36,37,38] These systems exploit special knowledge to solve difficult problems in many domains. Examples are: DENDRAL, mass-spectrum analysis; MYCIN, medical diagnosis; PROSPECTOR, mineral exploration; R1, VAX computer configuration; to name only a few. This chapter introduces general components of KBESs, describes their general features, describes two implementations, and provides references to several other systems. Chapter 3 discusses how the KBES approach has been used to develop an intelligent CAD tool, DAA.

2.1 General Components of a Knowledge-Based Expert System

The problem domains and features of the existing KBESs differ widely, but many have three components in common: a working memory, a rule memory, and a rule interpreter.

2.1.1 The working-memory component. The working memory is a collection of attribute-value pairs that describe the current situation. They resemble the data structures in conventional programming languages:

```
struct foo {
    <attribute 1> = <value 1>;
    ...
    <attribute n> = <value n>;
};
```

Some systems also represent goals and sub-goals as named attribute-value pairs in working memory.

2.1.2 The rule memory component. The rule memory is a collection of conditional statements that operate on elements stored in the working memory. The statements resemble the conditional statements of conventional programming languages:

```
IF:
    <antecedent 1>
    ...
    <antecedent n>

THEN:
    <consequence 1>
    ...
    <consequence m>
```

The rule memory is a collection of knowledge *chunks* about a particular problem domain. Most rule-based systems contain hundreds of rules that have been *painfully* extracted from months of interviewing experts. Acquiring knowledge from experts is difficult because although they are skillful at doing the task, they are not effective at explaining precisely how they do the task. To give a feel for the number of rules: MYCIN has about four-hundred fifty rules, R1 has about eight-hundred fifty rules (the new version XCON has three-thousand three-hundred rules),[39] and PROSPECTOR has about one-thousand six hundred rules. Systems like MYCIN have added debugged rules at an average rate of about two per week. This in no way accounts for the hundreds of rules that came and went in the debugging of the rule memory.

2.1.3 The rule interpreter component. The rule interpreter pattern matches the working-memory elements against the rule memory to decide what rules apply to the given situation. Some tool-kits allow a degree of certainty to be associated with each consequence that suggests the degree to which the consequence follows from the antecedents. Others allow a pair of certainties to be associated with each consequence that suggests both how sufficient the presence of the antecedents are for establishing the consequence and how sufficient the absence of the antecedents are for not establishing the consequence. Still others apply the consequences with absolute certainty if the antecedents are present in working memory. The selection process of rules can be data driven, goal driven, or a combination of data and goal driven.

A data-driven selection process looks through the rule memory for a rule whose antecedents match elements in the working memory. This is also called forward-chaining or antecedent reasoning. The consequences of the rule are applied, and the process is repeated until no more rules apply or until a rule explicitly stops the process.

A goal-driven selection process looks through the rule memory for a rule whose consequences can achieve the current goal. This is also called backward-chaining or consequent reasoning. If the antecedents of this rule match the elements in working memory, then the consequences are applied and the goal is satisfied. If the antecedents of this rule are not all present in working memory, then the missing antecedent replaces the current goal and the process is repeated until either all the sub-goals are satisfied or no more rules are applicable. If no more rules are applicable, the user can be queried for missing information to place in working memory. This backward-chaining reason also facilitates explanations of how the system had reached a particular conclusion and why it needed certain pieces of knowledge.

2.2 General Features of a Knowledge-Based Expert System

Separating expert knowledge from the reasoning mechanism implies several general features common to knowledge-based expert systems. The knowledge engineer can incrementally add new rules and refine old ones because the rules are chunks of domain knowledge and have minimal interaction with other rules in the rule memory. The rule interpreter can be queried about *why* it needs an additional piece of information and *how* it solved a problem by describing the goals and the rules it has used to solve a problem. A rule interpreter can be developed for aiding the acquisition of knowledge,[40] by checking its own rule memory for oddity, consistency and omitted areas. A rule interpreter can also be developed for aiding the instruction of knowledge,[41] by trying to determine what knowledge students seem to possess and how that knowledge corresponds to its rule memory. Finally, the same rule interpreter may be used by many rule memories and working memories to develop KBESs for different problem domains.

2.3 Example Knowledge-Based Expert Systems

During the past decade many KBESs have been developed for such purposes as interpretation, diagnosis, monitoring, prediction, planning and design. This section discusses two examples — a medical diagnostic system, MYCIN, and a system used to configure VAX computers, R1, and provides references to further readings in Tables 1 and 2. The commentary provides a description of the task and a discussion of the rule interpreter with its current states.

Table 1. A DECADE OF SYSTEMS

System	Domain
INTERNIST[42]	diagnosis in medicine
MYCIN[43]	diagnosis in medicine
PUFF[37]	diagnosis in medicine
GUIDON[44]	computer-aided instruction in medicine
VM[45]	measurement interpretation in medicine
SACON[46]	diagnosis in structural engineering
PROSPECTOR[47]	diagnosis in geology
DENDRAL[48]	mass-spectrum analysis in chemistry
SECHS[49]	organic chemistry
SYNCHEM[50]	chemistry
SADD[51]	electronics
EL[52]	circuit analysis
PALLADIO[34]	VLSI design
SOPHIE[53]	computer aided instruction in electronics
MOLGEN[54]	problem solving and planning in genetics
NEWTON[55]	problem solving and planning in mechanics
PECOS[56]	problem solving and planning in programming
DART[57]	diagnosis in computer faults
R1[58]	configuring VAX computers
XSEL[59]	computer sales person assistant
CONCHE[60]	knowledge acquisition
TEIRESIAS[40]	knowledge acquisition for MYCIN systems
HEARSAY-II[61]	speech recognition

Table 2. A DECADE OF TOOLS

System	Domain
AGE[62]	blackboard model tool-kit
EMYCIN[63]	MYCIN tool-kit
EXPERT[64]	diagnosis in medicine
HEARSAY-III[65]	HEARSAY-II tool-kit
KAS[66]	PROSPECTOR tool-kit
META-DENDRAL[67]	DENDRAL tool-kit
NEOMYCIN[68]	computer aided instruction in MYCIN
AMORD[69]	an EL tool-kit
OPS3[70]	an OPS family tool-kit
OPS5[28]	an OPS family tool-kit
ROSIE[71]	a RITA tool-kit

2.3.1 The MYCIN example. MYCIN is an interactive program for medical diagnosis. It produces diagnoses of infectious diseases, particularly blood infections and meningitis infections, and advises the physician on antibiotic therapies for treating those infectious diseases. During the consultation, which is conducted in a stylized form of English, the physician is asked only for patient history and laboratory test results. The physician may ask for an explanation of the diagnosis and the reason a laboratory test result is required.

MYCIN uses a goal-driven rule interpreter to scan a rule memory of about five hundred rules covering meningitis and blood infections. Each rule has a certainty factor that suggests the strength of the association or degree of confidence between the antecedents of the rules and the consequences of the rules. MYCIN has equaled the performance of nationally recognized experts in diagnosing and recommending therapy for meningitis and blood infections. However, it is not clinically used because the human factors of its interface do not save the doctors any time and its knowledge is limited to these two domains.

2.3.2 The R1 example. R1 is a program for configuring VAX computers. It organizes the customer selected components by location relative to other components or to the rooms in which the system will be housed, and it specifies the cabling required to connect pairs of components. This configurer is given only the set of components selected by a customer and produces the list of missing components, components in each CPU cabinet, components in each UNIBUS box, distribution panels in each cabinet, the floor layout, and cabling requirements.

R1 uses a data-driven rule interpreter to scan a rule memory of about eight hundred rules covering component configuration. These rules are written in the OPS5 programming language and use a direct association between antecedents and consequences. It takes about seventy five seconds on a VAX 11/780 to configure a typical order of about 90 components. R1 has equaled the performance of Digital Equipment Corporation experts in configuring VAX computers and is currently being used in a production mode.

2.4 KBES Summary

During the past decade many KBESs have been developed by researchers in artificial intelligence that assist experts in solving real-world problems such as interpretation, diagnosis, monitoring, prediction, planning and design. These systems differ from previous efforts in problem solving by effectively coping with the enormous search spaces of alternatives found in real-world problems. The key to success in KBESs is the ability to use domain-rich knowledge to recognize familiar patterns in the current problem state and act appropriately. Development of KBESs is aided by the separation of working memory, rule memory, and rule interpreter, which enables *how* and *why* questions to be asked. Researchers have developed many KBESs whose features have matured and grown into usable engineering tools. Most importantly, this tool is appropriate to the design domain because there are human experts available whose knowledge has been gained through experience and who can teach this knowledge through apprenticeship. Furthermore, the knowledge required to do the task is extensive and requires the type of organization provided by the KBES approach.

Chapter 3

DAA DEVELOPMENT

KBESs are generally developed in several stages. First, "book knowledge" of the problem is codified as a set of situation-action rules; interviews with experts then fill in knowledge gaps and refine current knowledge. Then, many example problems are given to the KBES, and experts closely examine and validate the results. Often, errors are found through the examples, and new rules are added to the system to correct the error situations.

This iterative process is necessary because experts are often unaware of exactly how they go about designing a chip and are inexperienced at articulating the procedure. Furthermore, the knowledge base is not an exact codification of the expert's knowledge, but a compilation of what is understood by the knowledge engineer.

This section discusses the development cycle of DAA, starting from the case-study interviews that led to the prototype system, followed by the acquisition interviews that have developed the DAA system, and concluding with an analysis of the acquired knowledge.

3.1 Methods to Acquire Expert Knowledge

The first step to extract knowledge that expert VLSI designers use to do allocation is to examine elicitation methods used by professionals. Extracting knowledge from experts is difficult because although they are experts at doing the task, they are likely to be novices at explaining precisely *how* they do the task. Though acquiring expert knowledge is a difficult problem, there are many elicitation procedures in psychology and cognitive science. A spectrum of detail, from broad-based to specific, can be elicited by using techniques that are unstructured, interviewer-structured, or interviewee-structured, depending on purpose and taste. This section discusses two elicitation procedures and a characterization of designer expertise.

A case-study is an interviewee-structured face-to-face session in which the expert is asked to explain how a task is done. This approach gives information about a few of the problems the expert has encountered, but not the complete set of possible problems. This may be because an expert "knows" what to do by recognizing a familiar pattern in the current situation and remembering what actions to take for that situation. Thus, the expert must be presented with a pattern to remember what to do. Acquiring knowledge from several experts by this approach yields a broad base of isolated pieces of information, which is a good method for initial knowledge acquisition.

An acquisition interview is an interviewer-structured face-to-face session presenting a specific problem and its solution, and asking the expert to point out where and why the solution went wrong. This approach can uncover the unspoken steps for doing a task. Acquiring knowledge from several experts by this approach fills in the gaps from previous knowledge acquisition attempts and provides a good method for verifying and testing the acquired knowledge.

Along with the many elicitation methods, there are three possible populations of designers to interview, each of which has its own strong and weak points:

- An inexperienced designer may not have a well formed method of doing design, but may be better at explaining the text book designs.
- A designer with a moderate amount of experience may be starting to form design methods based on what has worked in previous designs, but may have forgotten some of the reasons for the original decisions.

- An expert designer has the best formed design methods, but may only be able to explain the decisions when pushed for the reasoning.

If DAA is to model expert behavior, it is important for the acquisition interviews to use experts. On the other hand, for the case-study method, it would be informative to use designers from all three ranges of experience. Thus, all types of designers were used in the case-study interviews, while only expert designers were used for the acquisition interviews.

3.2 The Case-Study Interviews

After gathering current book knowledge about synthesis of the implementation design space,[6,4,5] we taped interviews with four designers and summarized the interviews. This section overviews those interviews, while the next section discusses the design of the prototype system.

We interviewed four designers of varied experience: one was a novice, two were moderately experienced, and one was an expert. The interviews, which lasted about an hour each, started with a determination of the designer's background, including years of experience, logic families used, and designs created. Most of the time was spent discussing the design process, with some time given to a discussion of the DAA system. The designers discussed the global picture, partitioning, selection and allocation tasks, each with sub-tasks or attributes. Our interview method was designed to allow the interviewees as much freedom as possible in generating ideas; we emphasized such questions as:

"What do you do next?"
and
"Could you elaborate?"

Using this interview method the designers found it hard to be specific about tasks and sub-tasks. However, by taping the interviews our interaction and involvement with the designers could encourage them to be specific.

The first interview served as an initial exploration of the interview technique. It was difficult to get the designer to talk about the specifics of the design process without adding a bias. Three more designers were interviewed in the next few days.

As expected, the case-study interviews gave information about a few of the problems the experts had encountered, but not the complete set of possible problems. When they did discuss details, they spoke about

Table 3. SUMMARY OF INTERVIEWS — PART 1

Interview	1	2	3	4
Background				
Level	novice	moderate	expert	moderate
TTL & ECL	yes	yes	yes	yes
NMOS	no	yes	yes	yes
CMOS & PMOS	no	no	yes	no
Industrial years	3	4	12	4
Total years	7	9	12	7
Global Picture				
Inputs	yes	yes	yes	yes
Outputs	yes	yes	yes	yes
Constraints	yes	yes	yes	yes
Functionality	yes	yes	yes	yes
Feasibility	yes	yes	yes	yes
Technology Independence	yes	no	yes	no
Partitioning				
Functions	yes	yes		
Communication	yes	yes	yes	yes
Style	yes			
Nodes			yes	
Ordering				
Constraints	1	1	1	1
Output, Input	2	2		
Regular Structures			2	2
Initial Style				
Parallel	yes	yes		yes
Old Designs	yes	yes		yes
Serial			yes	

recognizing a situation and taking the appropriate action. The case-study method gave background knowledge that provided a framework for the other knowledge acquisition methods, but was insufficient for building the complete DAA system.

Table 4. SUMMARY OF INTERVIEWS — PART 2

Interview	1	2	3	4
Allocation				
Clock Phases	1		2	
Operators	2	3		
Registers	3	2	3	
Data Paths	4	1	1	1
Control Logic	5	4	4	2
Iterating				
Constraint Violations	yes	yes	yes	yes
Global Improvements	yes			yes
Technology Increase	yes	yes	yes	yes
Functionality Decrease		yes	yes	yes
Constraints				
Speed	yes	yes	yes	yes
Area		yes	yes	yes
Power			yes	yes
Schedule			yes	
Cost	yes	yes	yes	
Drive	yes			yes
Width	yes	yes		
Bottom-Up				
After Global Picture	yes	yes	yes	yes

Tapes of the case-study interviews were replayed several times to construct a two-level classification of responses by each person. Each designer was given the information from his interview and asked to review it for accuracy and omissions. Tables 3 and 4 show the classifications, while the following sections explain the entries. A *yes* entry in the table shows the designer mentioned the item and said he *did* do it. Even though many entries are marked *yes*, they do not mean each designer did the same thing. A *no* entry in the table shows the designer mentioned the item and said he *didn't* do it. A missing table entry shows the designer never mentioned the item. A *numeric* entry shows the designer mentioned the item and ascribed an order to this item with relation to other items he mentioned.

3.2.1 The background of the designers. Four designers[72,73,74,75] of varied experience were interviewed: one was a novice, two were moderately experienced and one was an expert. The ratings were established by the amount of experience each designer had with MSI and LSI design techniques.

Designer-1 is a novice designer and home-computer hobbyist. He has been designing with TTL and microprocessor-based systems for about seven years. During the last three years he has worked for several small companies as a design engineer. He has designed memory systems, terminal interfaces, cassette tape recorder interfaces, and modem interfaces.

Designer-2 is a moderately experienced designer and home-computer hobbyist. He has been designing with TTL, ECL, and a little NMOS for nine years. During the last four years he has been working in industry and pursuing his doctorate. He has designed a bus controller for a disk drive and a Digital Equipment Corporation Q-Bus for a Motorola 68000 microprocessor.

Designer-3 is an expert industrial designer. He has been designing with TTL, ECL, NMOS, and CMOS for twelve years. He has designed a programmable controller, a four-bit microprocessor and is now managing a team designing a new microprocessor.

Designer-4 is a moderately experienced industrial designer. He has been designing with TTL, ECL, and a little NMOS for seven years, the last four years in industry. He has designed a high-resolution bit-mapped display and the registers and ALU of a 32-bit microprocessor.

3.2.2 Global picture of the design. The designers began with a high-level overview of the hardware, which listed inputs and outputs to the outside world, the functions the hardware should provide, and general constraints. They discussed design feasibility with considerations of the target technology. Two designers internalized the global picture in functions allowed by the target technology. Designer-4 said, "There is always a right logic family for a design problem."

3.2.3 Partitioning a design problem. The designers partitioned the global picture into smaller units to be dealt with separately. They emphasized minimizing connections among blocks. One designer also partitioned by groups that operated as parallel or serial units. Two designers put functions/operators of similar types together. One designer partioned by common multiplexer or communication bus.

3.2.4 Ordering the partitions within a design. Once the design was partitioned, the partitions were chosen for allocation in a decreasing order of difficulty or degree of constraint. Designers-1,2,3 reasoned that if the most difficult part could be designed, the rest of the design was feasible. Two designers worked in a grand sweep fashion: from output to input and input to output, from left to right, or from top to bottom. The other two designers continued by selecting regular structures next.

3.2.5 Initial style. Once a partition was selected for allocation, it was carried out either in parallel or in series. Designers-1,2,4 said, "A parallel design made thinking of the control logic much simpler," while designer-3 said, "A serial design minimized the design area." The constraints of the parallel design were examined for size violations to determine the parts to be serialized by adding data paths, registers, and control logic to the initial parallel design. The constraints of the serial design were examined for speed violations to determine the parts to be reimplemented in parallel. If the designers recognized a part of the design as similar to a part of a previous design, they used what they knew had worked in the past.†

3.2.6 Allocation. Within each partition, designers allocated clock phases, operators, registers, data paths, and control logic. The order was interesting because once registers and data paths were allocated, they were not changed. Designer-2 said, "The control was changed after the data paths because it was the hardest thing to think about and because it depended on a constant structure of data-path elements." Designer-3 chose registers by what values had to be latched between clock phases/cycles. Designers-2,3 picked data paths by a general style of multiplexer or bus, depending on the expected amount of traffic between the two points.

3.2.7 Iterating through a design's lifetime. The designers described the iteration process as a step-by-step refinement to meet violated constraints. They looked for a technology change to meet a constraint before making a design change. This could be as simple as finding a new chip in the TTL

---

† I feel this recognition of past designs is like the opening book moves for many of the chess playing computer programs.

data book or as complicated as a design-rule shrink. Three designers described giving up functionality to meet a constraint. Other design changes consisted of global improvements not recognized until the design neared completion. This suggests that the general choice of partitions and the initial design style selections approached optimum. This is an important point in the later development of the DAA system because designers do not seem to use backtracking in their designs. The designers were willing to make local iterations and improvements to clean up the design, but once they allocated a piece of hardware they did not retract it.

3.2.8 Use of design constraints. The whole design process seems to be a large multi-variable constraint problem. Designer-4 summed it up best by saying,

> "An engineer's training teaches when
> constraints can be swept under the rug."

The relative importance of constraints is application dependent. The designers mentioned the constraints of speed, area, power, schedule, cost, drive capabilities, and bit width.

3.2.9 Top down to bottom up designs. The designers started with a top down approach, and then chose a bottom up approach for designing the most constrained partition. The switch to a bottom up approach seemed to occur after the global picture was completed. Designers-1,3 iterated using both the top down and bottom up paradigm. Designer-4 proceeded mostly top down, possibly because of his hierarchically oriented design tool, DRAW,[76] and software training. The other designer proceeded without a particular order.

3.3 Prototype System

Even though many details were missing, enough book knowledge had been gathered to put together a prototype version of the DAA system using the OPS5[28] KBES writing system. The initial knowledge in the system was codified from the algorithms of the current CMU/DA allocator[6] and the interviews discussed above. Throughout the development of the DAA system the experts interacted with the knowledge engineer, not the rules themselves. While the DAA system was far from perfect at this point, it stimulated further elicitation sessions with expert designers. This section discusses the flow of control in the prototype system and how the KBES approach formulates the problem. Chapters 4 and 5 present the complete working-memory representations and a detailed analysis of the subtasks after the knowledge base stabilized.

The DAA starts with a data flow representation extracted from the algorithmic description. This representation resembles the internal description used by most optimizing compilers, but computer programs manipulate it more easily, and it is felt to be less sensitive when the same algorithm appears in a variety of writing styles. The DAA produces a technology-independent hardware network description. This description is composed of modules, ports, links, and a symbolic microcode. The modules can be registers, operators, memories, and buses or multiplexers with input, output, and bidirectional ports. The ports are connected by links and are controlled by the symbolic microcode.

The DAA uses a set of temporally ordered subtasks to perform the synthesis task. It begins by allocating the base variable storage elements — constants, architectural registers, and memories with their input, output and address registers — to hardware modules and ports. Then a data-flow BEGIN/END block, or VT-body,† is picked, and the synthesis operation assigns minimum delay information to develop a parallel design. Next, it maps all data-flow operator outputs not bound to base-variable storage elements to register modules. Last, it maps each data-flow operator, with its inputs and outputs to modules, ports, and links. In doing so, the DAA avoids multiple assignments of hardware links; it supplies multiplexers where necessary. These last two mapping steps place the algorithmic description in a uniform notation for the expert analysis phase that follows.

The expert-analysis subtask first removes registers from those data-flow outputs where the sources of the data-flow operator are stable. Operators are combined to create ALUs, according to cost and partitioning information across the allocated design. The DAA also examines the possibility of sharing non-architectural registers. Where possible, it performs increment, decrement and shift operations in existing registers. Where appropriate, it places registers, memories, and ALUs on buses. Throughout this subtask, constraint violations require trade-offs between the number of modules and the partitioning of control steps. The process is repeated for the next data-flow BEGIN/END block.

---

† VT-bodies are fully described in Chapter 4.

The DAA is implemented as a production system via the OPS5 KBES writing system. The KBES tool is based on the premise that humans solve problems by recognizing familiar patterns and by applying their knowledge in the current situation. The tool formulated a problem by using three major components: a working memory, a rule memory, and a rule interpreter. Chapter 2 describes KBESs and provides additional examples.

3.3.1 The working memory. The working memory is a collection of attribute-value pairs that describe the current situation. They resemble records or structures in conventional programming languages:

> literalize module
>   id: adder.0
>   type: operator
>   atype: two's complement
>   bit-left: 17
>   bit-right: 0
>   attribute: +

This working-memory element describes an operator module *adder.0*, which can perform two's complement addition on 18 bits of binary data.

3.3.2 The rule memory. The rule memory is a collection of conditional statements that operate on elements stored in the working memory. The statements resemble the conditional statements of conventional programming languages:

IF:
   the most current active context is to create a link
   and the link should go from a source port to a destination port
   and the module of the source port is not a multiplexer
   and there is a link from another module to the same destination port
   and this other module is not a multiplexer

THEN:
   create a multiplexer module
   and connect the multiplexer to the destination port
   and connect the source port and destination port link to the multiplexer
   and move the other link from the destination port to the multiplexer

This rule recognizes situations in which a multiplexer needs to be created to connect one port to another.

Each subtask in the DAA is associated with a set of rules for carrying out the subtask. An example of a rule for the fourth subtask appeared above. Most of the rules, like the example above, define situations in which a partial design should be extended in some particular way. These rules enable the DAA to synthesize an acceptable design by determining, at each step, whether a certain design extension respects constraints.

3.3.3 The rule interpreter. The rule interpreter pattern matches the working-memory elements against the rule memory, to decide what rules apply to the given situation. The rule-selection process is data driven; the rule interpreter looks through the rule memory for a rule whose antecedents match elements in the working memory. This is also called forward chaining or antecedent reasoning. The consequences of the rule are applied, and the process is repeated until no more rules apply or until a rule explicitly stops the process. If more than one rule applies, the rule dealing with the most current working memory is selected first. If multiple rules are still applicable, the most specific rule is selected. This selection mimics following a train of thought, as far as possible, and uses special-case knowledge before general-purpose knowledge.

The separation of expert knowledge from the reasoning mechanism makes the incremental addition of new rules and the refinement of old ones easy because the rules have minimal interaction with one another. By using a KBES approach, DAA uses a *weak method*[77] called *match* in place of extensive backtracking. DAA uses match to explore the space of possible designs by extending a partial design from an initial state to a final state *without* any backtracking. DAA proceeds through its major tasks in the same order for each problem; it never varies the order and it never backs up in any problem. This means that at any intermediate state, DAA can determine how to extend the design to achieve an acceptable result. The advantages of using the KBES approach are the ease of validating the knowledge gathered from interviews with experts, the ease of incrementally adding to the knowledge base, the ability to query that knowledge during the design task, and the replacement of extensive backtracking by domain-specific knowledge techniques. The disadvantage is the difficulty of extracting knowledge from the experts.

3.4 Knowledge Acquisition Interviews

The prototype DAA system had about 70 rules and could design a MOS Technology Incorporated MCS6502 microcomputer in about three hours of VAX 11/750 CPU time. We asked many expert designers at INTEL and AT&T Bell Laboratories to critique the design by explaining what was wrong, why it was wrong, and how to fix it. After each critique, rules were modified, new rules were added, and the MCS6502 was re-designed. Based on the critiques, the development DAA system now has over 300 rules, and has designed a much better MCS6502 microcomputer in about five hours of VAX 11/750 CPU time. In retrospect, clearly much of what we learned was common-sense design knowledge, the same things human designers learn through apprenticeship. The DAA has undergone many improvements and produced many designs of the MCS6502 microcomputer. Below, we illustrate a few of these changes.

Each knowledge acquisition interview began by giving the designer a drawing of the design with a sheet of clear plastic over it. Before the designer started the critique, pieces of cardboard were placed over the design. As the designer proceeded, a piece of cardboard had to be lifted, the plastic written on, and covered by new plastic to correct the design. This provided a complete record of where the designer was focusing attention and what was corrected. The designers found this elicitation procedure compatible with their normal spatial mode of operation. The first prototype DAA system was used to produce the design summarized in Column 1 of Table 5. Each row shows the bits of the specified operator or register type found in the design.† The expert criticism is summarized in four points:

- Operators of different types and sizes should be combined into ALUs.
- One-bit operators within the same block should not be combined, because multiplexers are more expensive than most one-bit modules.
- Registers should increment, decrement, and shift their values internally, where possible.
- Temporary registers to the controller should be eliminated, and one latched register should be placed in front of the controller.

The rules were changed to produce the design summarized in Table 5, Column 2, and illustrated in Figure 1. To produce this design, partitioning information was added, based on connectivity of data paths and similarity of operators among blocks. The details of the partition and cost estimators is given in Chapter 5. This simplified the decision about which modules to

---

† A figure is not provided because it is totally inscrutable.

Table 5. MCS6502 — THREE DESIGNS

Designs	1	2	3
And	20	20	20
Cmp	177	1	1
Minus	64	0	0
Or	9	9	9
Not	21	21	21
Plus	540	0	0
Shifts	35	1	1
Xor	9	9	9
Alu	0	35	35
Dreg	450	281	210
Treg	1227	0	62
Mux In	2122	2657	473
Mux Out	293	377	84
Bus In	0	0	769
Bus Out	0	0	210

combine when hardware operators are shared among abstract operations detailed in the algorithmic description. Rules were also added to combine modules of different sizes and types. As Column 2 shows, the ALU number increased, decreasing the plus, minus, shift, and compare numbers. Rules were also added to decrease the amount of temporary register storage.

Figure 1 shows the eight-bit data paths of the MCS6502. The one-bit and 16-bit data paths were omitted for clarity. Each of the symbols represents a module. The circles are single function ALU modules to AND, SHIFT, NOT, XOR, and OR data, the small trapezoids are multiplexers that gate one of their inputs to the output, the small rectangles are registers, the large rectangle is the memory, the large trapezoid is a multi-function ALU, and each of the lines represents a link between the modules. Where the links join with the modules, a port is defined. An obvious problem, pointed out by our experts, was overuse of multiplexers. They suggested ways of distributing the multiplexer hardware to form buses.

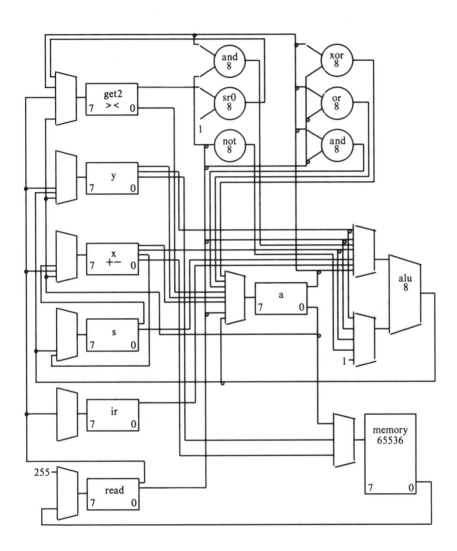

Figure 1. MCS6502 — 8 BIT MUX DATA PATHS

The rules were changed in response to these critiques, resulting in the design of Table 5, Column 3, and Figure 2. To produce this design, rules were added to recognize when a multiplexer should be converted into a bus and how to share that bus with other distributed multiplexers. In addition, new rules decreased the amount of declared register storage. Specifically,

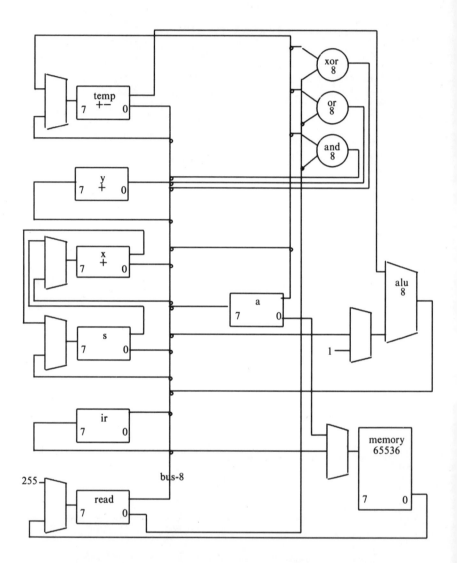

Figure 2. MCS6502 — 8 BIT BUS DATA PATHS

registers were not needed to multiplex information into the data flow BEGIN/END blocks. As Column 3 shows, the multiplexer numbers decreased, increasing the bus numbers. The declared register number also dropped.

Though this design was acceptable to our experts, it was not perfect. Further changes led to improvements such as multiple buses of different widths. However, these changes did not affect the MCS6502 because it did not require multiple buses. This brings up an interesting point about expert systems: they are never totally finished. Like human designers, the DAA becomes a better designer as its rule memory expands. Until all possible world knowledge about designing microprocessors has been codified in the DAA's rules, there will always be room for improvement in its designs. However, for the data representation, knowledge representation, and System/370 experiment discussions in Chapters 4, 5, and 6, the rules were held constant.

3.5 Analysis of Knowledge

The last step in extracting knowledge is to examine it and determine what has been learned. One merit of codifying knowledge in a KBES is that it can be easily quantified and qualified. Chapters 4 and 5 provide analysis by examining the representations and functional knowledge in the DAA. This section provides additional analysis by examining both the extent to which each rule embodies domain knowledge and the type of each rule. The analysis includes the selection criteria of types, the English translation of several rules, and a brief discussion of the translated rules. These translations must be taken with a grain of salt because rules do not stand by themselves, but are part of a larger network of rules connected by the working memory and the inference engine.

Table 6. RULES BY KNOWLEDGE TYPE

Domain Specific			Domain Independent		
Type	Rules	Firing	Type	Rules	Firing
Design	151	3768	Cleanup	33	1586
Context	18	1562	Input	9	64
Setup	70	70	List	14	1290
			Test	19	203
Total	239	5400	Total	75	3143
Total DAA	314	8543			

The rules in DAA either contain domain specific or domain independent knowledge. Knowledge directly related to the implementation design task is domain specific, while knowledge of a more general type is domain independent. An example of domain-specific knowledge is how to make particular design decisions and an example of domain-independent knowledge is how to count things. Distributed between these two categories DAA contains seven distinct rule types. Rules that extend the partial design, control the context or focus of attention, and maintain the technology-independent database are domain specific. Rules that remove unneeded working-memory elements, transform the input description, simulate list and set operations, and simulate procedure calls or calculations in antecedents are domain independent.

Table 6 summarizes these categories by listing the number of rules in each category and the number of rule firings for the complete design of the SCF3 processor. The structured control flow processor,[78] SCF3, was chosen partly because it is slightly smaller than the MCS6502 design (two hours of VAX 11/780 CPU time), but mostly because it is being carried through the CMU/DA system to produce a working silicon chip. Table 6 shows that three quarters of the rules and two thirds of the rule firings are domain specific. Most of the domain specific rules and firings are used to extend the partial design (See Row Design in Table 6). The remaining rules and rule firings are domain independent or overhead rules. Most of these rules and firings are used to clean up unneeded working-memory elements and simulate list and set operations (See Rows Cleanup and List in Table 6).

3.5.1 Rules that extend the design. This category of rules can be distinguished from other rules in DAA by their modification action to the design.

  IF:
    the most current active context is bus allocation
    and there is a link from a module to a bus
    and there is a link from the bus to a multiplexer
    and there is a link from the module to the multiplexer
  THEN:
    check to see if the bus is idle during the control step needed
      to multiplex the values

Figure 3. MODULE-TO-MUX-AND-BUS-WITH-BUS-TO-MUX

They either make design decision directly, or use a test rule to make design decision for them. The rule in Figure 3 extends the partial design by recognizing the need to use a bus to make a connection and creates a context for a test rule to see if the bus is idle. If the bus is idle the connections are moved from the multiplexer to the bus. This rule uses a test rule to complete a match. If OPS5 could call procedures during the match cycle, this rule and the appropriate test rule would have been merged. The rule in Figure 4 extends the partial design by recognizing the need for an incrementer and assigns the increment function to the correct register. This is an example of a rule that directly extends the design.

IF:
>the most current active context is to allocate value trace operators
>and there is a value trace + operator
>and one input to the + operator is a register
>and the other input is the constant one
>and the output of the + operator is the input of a bit read operator
>and the output of the bit read is the full width of the register feeding the + operator

THEN:
>add the increment attribute to the register

Figure 4. ASSIGN-INC

It could have been written as a number of separate rules using constants to link the type of bound and unbound modules together. This would have increased the number of rules, decreased the CPU time to run the system, and increased the difficulty of modifying the rule base. This general style of writing rules is less efficient in computer time, but more efficient in rule-base modification time.

IF:
>   the most current active context is fold allocation
>   and there is an unbound module whose width is greater than 1
>   and there is a database operator for this module whose group is either ARITHMETIC, RELATIONAL, SHIFT, or ALU
>   and there is a second database operator whose group is either ARITHMETIC, RELATIONAL, SHIFT, or ALU
>   and there is a bound module whose operator is in the set of the second database operator with the same width as the first module

THEN:
>   check to see if the proximity and cost factors show they should be combined

Figure 5. BIND-ARITHMETIC-OPERATORS-SAME-SIZE

3.5.2 Rules that control the context or focus attention. This category of rules can be separated from other rules in DAA because they only manipulate the context status. By manipulating active, pending, next and not-satisfied status codes, these rules control the focus of the design task. They also provide the ability to sequence a list of tasks.

IF:
>   there is a request to allocate a value trace body

THEN:
>   remove the request
>   and make the most current active context to allocate temporary variables
>   and make the next pending context to allocate control steps
>   and make the next pending context to allocate operators
>   and make the next pending context to fold allocation

Figure 6. ALLOCATION

The rule in Figure 6 generates a number of pending contexts and thus controls the focus of attention of DAA. The pending contexts are modified into active contexts when there are no other rules to fire in the currently active context. This helps minimize the set of rules that could fire and decreases the number of internal match states in OPS5. This is an example of one of six rules that set up a list of subtasks. The rule in Figure 7 modifies a waiting context to an active state and thus controls the focus of

IF:
> the most current context is a next context

THEN:
> make it the most current active context

Figure 7. NEXT

attention of DAA. About ten percent of the total rule firings are composed of this rule. At first glance it seems to be an unnecessary rule, and indeed it is! However, using it reduces the number of rules that can fire, which decreases the total CPU time in OPS5. The design time for the MCS6502 was reduced by four percent. This rule should be removed and the other rules modified if parallelism is measured on DAA. This is an example of one of three rules that monitor the active context and change context if appropriate, thus limiting the set of active rules.

3.5.3 Rules that manage the technology-independent database. The rules in this category are distinguished from other rules in DAA by their default database creation actions.

IF:
> the most current active context is to declare variable allocation
> and there is no database operator PLUS

THEN:
> make a database operator with hardware opcode PLUS,
>   control step delay 1, value trace opcode +, 0 modules
>   currently allocated, a maximum of 999 modules, and of group
>   ARITHMETIC

Figure 8. DEFAULT-PLUS

The rules in Figures 8 and 9 set up the technology-independent database. They fire once if the user has not provided the appropriate constraint value. Constraint information is kept in working memory and used by the rules as bound variables. This technique allows easy change of target technologies and individual fine tuning of design constraints. These are examples of rules that propagate default constraints through working memory to all the design rules.

IF:
>the most current active context is to declare variable allocation
>and there is no database fold parameter for ARITHMETIC types

THEN:
>make a database fold parameter with type ARITHMETIC,
>proximity and cost values of 0.500

Figure 9. DEFAULT-FOLD-ARITHMETIC

3.5.4 Rules that remove unneeded working-memory elements. This category of rules can be separated from other rules in DAA by observing that their actions are only to clean up the working memory.

IF:
>the most current active context is to clean working memory
>associated with value trace body N
>and there is an outnode associated with value trace body N

THEN:
>then remove it

Figure 10. CLEAN-OUTNODES

The rule in Figure 10 removes unnecessary working-memory elements, thus getting rid of unneeded details of the design. This domain-independent rule along with seven others reduces the size of working memory to the current VT-body and the hardware that has been allocated. Without these rules only small designs can run through DAA. This is an example of a domain-independent rule that aids DAA in execution, but does not perform any design task.

IF:
>the most current active context is to declare variable allocation
>and no other rule can fire

THEN:
>then remove the active context

Figure 11. DECLARED-VARIABLE-ALLOCATION-DONE

The rule in Figure 11 removes a currently active context that has been completely satisfied. This rule and several like it do not contribute to the design domain, but are simply aids in debugging the KBES. They aid debugging because there is a demon rule that dumps working memory to a file and lets the user interact with the KBES if an active context cannot be satisfied by any rule. This is an example of a domain-independent rule that helps knowledge engineers debug DAA.

3.5.5 Rules that transform the input description. This category of rules is similar to the rules that extend the design except they manipulate the input description. They either directly modify the input description, or use a test rule to make the change for them.

IF:
    the most current active context is to allocate temporary variables
    and there is a value trace operator of type +, −, or *
    and its output is bit read
    and the offset to the bit read is zero
    and the bit read produces fewer bits than does the value trace
        operator
THEN:
    make a trim record of the number of bits from the bit read

Figure 12. TRIM-ADD-SUB-MUL-OUTPUT-ZERO-NIL-TRIM

The rule in Figure 12 transforms the input VT description to a better description by minimizing the size of addition, subtraction and multiplication operators. This rule and a rule similar to it will repeatedly fire until the rule in Figure 13 finally limits the operator output size to the largest size required as input by any other operator. This is an example of a domain-independent rule that could have been added to the VT transformation package.

IF:
    there is a trim record for an outnode
    and the outnode has a larger number of bits than does the trim
        record
THEN:
    set the number of bits for the outnode equal to the trim record

Figure 13. TRIM-OUTPUT

The rule in Figure 13 waits until there are no more modifications of the trim record created by the rule in Figure 12 and then modifies the VT operator. This is an example of a rule associated not with the task of extending the design, but with modifying the input specification.

3.5.6 Rules that simulate set and list operations. This category of rules can be separated from other rules in DAA because it modifies the list working-memory element.

>   *IF:*
>   >   the most current active context is to step through a list
>   >   and create a new active context with the first object from the list
>   >   and there is an object on the list
>
>   *THEN:*
>   >   shift the first element off the list
>   >   and make a new active context with the first element of the list
>   >   >   as its object

Figure 14. NEXT-LIST-1

The rule in Figure 14 steps through a list of working-memory elements and creates a new goal for every element in the list. For example, this rule sequentially assigns control-step information to a list of VT operators.

>   *IF:*
>   >   the most current active context is to step through a list
>   >   and create a new active context with the first object from the list
>   >   and there not is an object on the list
>
>   *THEN:*
>   >   remove the active goal of next list

Figure 15. END-LIST-1

The rule in Figure 14 will stay active until the rule in Figure 15 recognizes there is nothing more to do and removes it. The usage of these rules increases for bigger designs because larger designs have longer lists to manage. These rules are examples of domain-independent rules that simulate set operations on a list of working-memory elements.

3.5.7 Rules that simulate procedure calls and calculations for antecedents. This category of rules can be distinguished from other rules in DAA because they only test the results of a procedure call or computation of a previous rule. The rule in Figure 16 combines two registers if the result of a layout estimation shows that they would be placed within a database minimum value of one another. If the result of a stability calculation shows that a register is required between control steps, the rule in Figure 17 marks it not stable. These are examples of domain-

IF:
>the most current active context is to fold two registers together with a given cost and proximity value
>and the proximity value exceeds the minimum required proximity value

THEN:
>combine the two registers

Figure 16. FOLD-REGISTER-PROX

independent test rules because they extend the ability of OPS5 to do procedure calls or calculations in the antecedent part of the rule. The domain-dependent part of the knowledge is provided by the rule that activated them.

IF:
>the most current active context is check the stability of a register
>and the register is not stable

THEN:
>then mark the register not stable

Figure 17. NOT-STABLE-REGISTERS

3.6 Development Summary

Chapter 3 has overviewed the development cycle of the DAA system. It has discussed the case-study interviews, which led to the prototype system, and the acquisition interviews that have further developed the system. Using a KBES approach has allowed the incremental addition of modular knowledge and queries about that knowledge during the design task.

During the case-study interviews the designers discussed the global picture, partitioning, selection, and allocation tasks. They began with a high-level overview of the hardware, which listed inputs and outputs to the outside world, the functions the hardware should provide, general constraints, and design feasibility with consideration of the target technology. They generally partitioned the global picture into smaller blocks and emphasized minimizing connections among blocks, selecting blocks that operated as parallel or serial units, and grouping according to similarity of function. Partitions were chosen for allocation in a decreasing order of difficulty or degree of constraint. The designers reasoned that if the most difficult part could be designed, the rest of the design was feasible.

Once a partition was selected for allocation, it was carried out either in parallel or in series. A parallel design made thinking of the control logic much simpler, while a serial design minimized the design area. The constraints of the parallel design were examined for size violations to determine the parts to be serialized by adding data paths, registers, and control logic to the initial parallel design. The constraints of the serial design were examined for speed violations to determine the parts to be reimplemented in parallel. If the designers recognized a part of the design as similar to a part of a previous design, they used what they knew had worked in the past. Within each partition, designers allocated clock phases, operators, registers, data paths, and control logic. The order was interesting because once registers and data paths were allocated, they were not changed. The control was changed because it was the hardest thing to think about and because it depended on a constant structure for the data-path elements.

The designers described the iteration process as a step-by-step refinement to meet violated constraints. They looked for a technology change to meet a constraint before making a design change. This could be as simple as finding a new chip in the TTL data book or as complicated as a design-rule shrink. Next, they would sacrifice functionality to meet a constraint. The relative importance of constraints is application dependent. The designers mentioned the constraints of speed, area, power, schedule, cost, drive capabilities, and bit width. Other design changes consisted of global improvements not recognized until the design neared completion. This suggests that the general choice of partitions and the initial design style selections approached optimum and that designers do not seem to use much backtracking in their designs.

A prototype system was built using the knowledge from the interviews and available book knowledge. The prototype system has been transformed into a development system through hundreds of hours of knowledge acquisition interviews with expert VLSI designers. In retrospect, much of what we learned was common-sense design knowledge, the same things human designers learn through apprenticeship. This learning process is not complete, nor will it ever be complete. Like human designers, the DAA becomes a better designer as its rule memory expands. Until all possible world knowledge about designing microprocessors has been codified in the DAA's rules, there will always be room for improvement in its designs.

Chapter 4

DAA REPRESENTATIONS

The DAA synthesizes a technology-independent representation of memories, registers, operators, data-paths and timing signals from an algorithmic description of a VLSI system. The two major components of any program are data and control structures. The DAA decomposes the problem into three data-structure representations and four control tasks. The representations are defined by the algorithmic-description language, VT, the constraints placed on the design, and the technology-independent-hardware-network description language, SCS. The tasks involve allocating global-storage elements, timing information, local-storage and processing elements, applying global improvements, and routing connections. They transform the VT representation and constraints into SCS using partition and cost estimators to provide abstract high-level floor-planning information. This problem division has been identified by designers and used to limit the complexity of the task of choosing implementations as discussed in Chapter 3. This chapter discusses the three representations and the bookkeeping information used by the DAA. Chapter 5 outlines the knowledge used by the DAA to design VLSI systems.

4.1 Algorithmic Representation

In the CMU/DA system, the input description is written in ISPS,[16] a hardware-description language that is similar in many ways to programming languages such as ALGOL or Pascal. An ISPS description consists of a series of declarations of *entities*. Some of these are simple *carriers*, which hold the data being manipulated, similar to variables in a programming language. Other entities are procedures or functions much like the procedures or functions of a programming language. These define how the data is manipulated. The procedures are specified by data *operators* that do arithmetic, logical, relational, and shift operations on the data, and by sequential, parallel, and conditional constructs that show how the data operators are combined.

```
                Example :=
                Begin

                ** Storage.Declaration **

                cpage\current.page<0:4>,
                i\instruction<0:11>,
                        pb\page.0.bit<>              := i<4>,
                        pa\page.address<0:6>         := i<5:11>

                **Address.Calculation**

                Global eadd\effective.address<0:11> :=
                        Begin
                        Decode pb =>
                                Begin
                                0 := eadd ='00000 @ pa,
                                1 := eadd = cpage @ pa
                                End
                        End
        End
```

Figure 18. SAMPLE ISPS DESCRIPTION

The ISPS description is considered an algorithmic description, not a representation of the structure of the implementation. Carriers do not necessarily represent individual registers. A carrier that simply holds temporary values, for example, might be implemented by a bus, which

Section 4.1   Algorithmic Representation   43

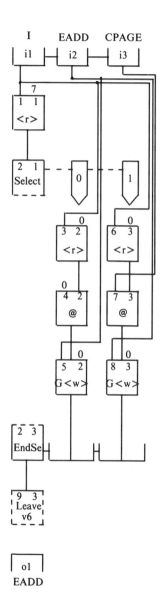

Figure 19. SAMPLE VT

simply transfers values to the next stage without storing them, or several carriers may share the same physical register. On the other hand, several instances of the same logical entity, such as an accumulator or program counter, might be required at the same time in a highly parallel implementation such as a pipelined machine. Here it would take several registers to implement the same carrier.

A simple ISPS example description is shown in Figure 18. This ISPS fragment is from the description of the Digital Equipment Corporation PDP-8. It first defines the current page and the instruction carriers. It then labels specific fields of the instruction carrier as the page-zero bit and page-address carriers. The last part of the ISPS fragment decodes the page-zero bit of the instruction carrier and sets the effective-address carrier equal to the page-address carrier concatenated with either zero or the current-page carrier.

The ISPS description is compiled into a VT[17,18] data-flow representation, which makes it easier to recognize and implement design decisions by synthesis programs. The VT is a directed acyclic graph, DAG, similar in nature to those used in optimizing compilers, with the addition of control constructs to allow conditionals and subroutines in the VT. The nodes in this graph are called *operators* and correspond to operations that take certain values as input and produce new values as output. They are translations of the ISPS unary and binary operations, operations that change or access fields in words or words in arrays, control operations such as procedure or block invocation, and conditional branches. The arcs connecting the nodes are called *outnodes* and represent the generation or use of data values. They are translations of the ISPS carriers and the temporary carriers needed to pass results from one operator to another. The graph is partitioned into subgraphs called *VT-bodies*, corresponding to a set of operations that can be evoked, entered, or left as a unit. These subgraphs are translations of ISPS procedures, labeled blocks and loops.

The VT fragment in Figure 19 shows the same decoding loop as the ISPS description in Figure 18. Each of the blocks represents a VT *operator*. For example, the operator @, called a concatenate, produces the concatenation of its input values. The operator <r>, called a *bit read*, produces the subfield of its left input that begins with the bit specified by its right input, and continues to the left for the width specified by the output. The operator G<w>, called a *global bit write*, produces the left input, with the middle input overwriting it, starting at the bit position specified by the right input. The compound operator consisting of the

Section 4.1   *Algorithmic Representation*   45

Table 7. VTBODY

Attribute	Values
vtbody	
id	id number
type	s=slist, r=carrier, m=mapping, v=vtbody
parent	id of vtbody declared in
entity-flag	vector: process, global, main, map, vtbody, slist-head, bit, word, external, repeat, multiplied, divided and loop
word-left	starting word
word-right	ending word
bit-left	starting bit
bit-right	ending bit
isp-name	from isps compiler
inputs	name of input outnode list
operators	name of operator list
outputs	name of output outnode list
calls	name of vtbody, or slist we call list
opcalls	name of operator list that call, enter, leave, restart, or resume this vtbody
map-id	mapping onto id
map-word-scale	mapping scaling factor
map-word-offset	mapping word offset from lowest numbered word in target
map-bit-offset	mapping bit offset from right most bit of the target mapped word
attributes	name of attribute list
qualifier-trees	name of qualifier tree
control-steps-assigned	control steps have been assigned
vt	vtbody number

blocks *SELECT* and *ENDSEL*, and the items connected to them by dashed lines, is called a *SELECT*. It is used in the VT to implement the DECODE and IF ... THEN ISPS constructs. In this example, it implements the decoding of the instruction register from Figure 18. Alternate actions are shown side-by-side with the value of the selector required for each shown at the top. The result of the chosen action is then passed to the ENDSEL where it continues to the LEAVE operator. The LEAVE operator

Table 8. OUTNODE AND OPERATOR

Attribute	Values
outnode	
id	id number
type	f=formal parameter, i=vtbody input, c=constant o=vtbody output, p=value from operator node
carrier-id	initially from isp
bits	number of bits
isp-name	name from isps compiler
value	value if a constant
vt	vtbody number
operator	
id	id number
opcode	operator
type	TC, OC, SM, US (from ISPS)
call	name of vtbody or slist we call, pstart restart, resume, or leave
inputs	name of outnode input list
outputs	name of outnode output list
attributes	name of attribute list
qualifier-trees	name of qualifier tree
parent	name of parent vtbody
branches	name of branch list for select and diverge
control-step-begin	starting control step number
control-step-end	ending control step number
vt	vtbody number

exits the VT-body and passes its input back to the calling VT-body.

For each design task the DAA is given a working-memory representation of the VT, shown in Tables 7, 8, and 9. The first column provides the name of the attribute, while the second column describes the possible values. An attribute that requires a variable-length list of values is either represented as a vector, denoted by *vector:*, or as the name of the list working-memory element that contains the values. This inconsistency is brought about because OPS5 can represent only one vector quantity in each

Section 4.1   *Algorithmic Representation*   47

Table 9. BRANCHES, LISTS AND TREES

Attribute	Values
branch	
id	id number
type	b=select, d=diverge
parent	parent operator
activations	name of list of activations for selects
operators	name of list of operators
inputs	name of list of inputs for selects
qualifier-trees	name of qualifier tree
vt	vtbody number
lists	
id	id number
list	vector: list elements
vt	vtbody number
trees	
id	id number
list	vector: list elements
tree	name of next tree
vt	vtbody number

working-memory element. Furthermore, the non-hierarchical limitation of the OPS5 working-memory representation increases the time required to match the rules because the graph is represented as pointer attributes to working-memory elements. The working-memory list of the ISPS and VT decoding loop in Figures 18 and 19 is given in Appendix A.

4.2 Technology Database and Constraints

The second input representation provides technology-sensitive information and design constraints. The information is referred to as technology sensitive rather than technology independent because it is sensitive to a particular technology rather than independent of technology considerations. The technology database provides the name translation between VT and SCS operators, and the number of micro-control steps

Table 10. USER PARAMETERS

Attribute	Values
db-operator	
hw-opcode	hardware operator code
control-step-delay	amount of delay
vt-opcode	value trace operator code
allocated	how many have I allocated so far
maximum	how many can I allocate
group	what group do the operators belong to
max-delay-per-control-step	
delay	number of delay units per control step
fold	
type	ARITHMETIC, LOGICAL, REGISTER
prox	proximity
cost	cost

required for the operation.† It also provides constraints on the number of each module the DAA may use in the design, thus providing rough limits on area and power. No constraints are given for speed because the DAA starts with the maximally parallel design, only serializing it as it hits area and power constraints. The technology database provides information about how many micro-control steps comprise a full-control step. This information is given in the working-memory element *max-delay-per-control-step*. The last technology-sensitive inputs are the proximity and cost thresholds for arithmetic, logical and register modules. They regulate how close a fit is required between two modules before they will be combined. Their use is further explained in Sections 5.4.2 and 5.4.4, which describe register and module allocation. Table 10 shows the working-

---

† The notion of micro-control steps is further explained in the Section 5.3.2, which describes operator assignment to control steps.

Table 11. DEFAULT DB-OPERATOR — PART 1

Hardware Opcode	Delay	VT Opcode	Max	Group
PLUS	1	+	999	ARITHMETIC
MINUS	1	−	999	ARITHMETIC
MULT	1	*	999	ARITHMETIC
DIV	1	/	999	ARITHMETIC
MOD	1	MOD	999	ARITHMETIC
UMINUS	1	−−	999	ARITHMETIC
AND	1	AND	999	LOGICAL
EQV	1	EQV	999	LOGICAL
OR	1	OR	999	LOGICAL
XOR	1	XOR	999	LOGICAL
NOT	1	NOT	999	LOGICAL
EQL	1	EQL	999	RELATIONAL
NEQ	1	NEQ	999	RELATIONAL
LSS	1	LSS	999	RELATIONAL
LEQ	1	LEQ	999	RELATIONAL
GEQ	1	GEQ	999	RELATIONAL
GTR	1	GTR	999	RELATIONAL
TST	1	TST	999	RELATIONAL
SR0	1	SR0	999	SHIFT
SR1	1	SR1	999	SHIFT
SRD	1	SRD	999	SHIFT
SRR	1	SRR	999	SHIFT
SL0	1	SL0	999	SHIFT
SL1	1	SL1	999	SHIFT
SLD	1	SLD	999	SHIFT
SLR	1	SLR	999	SHIFT
SLI	1	SLI	999	SHIFT
SRI	1	SRI	999	SHIFT

Table 12. DEFAULT DB-OPERATOR — PART 2

Hardware Opcode	Delay	VT Opcode	Max	Group
0PAD	0	PAD0	999	WIRING
SPAD	0	PADS	999	WIRING
FREAD	0	BIT-R	999	WIRING
FWRITE	0	BIT-W	999	WIRING
AREAD	−1	WORD-R	999	WIRING
AWRITE	−1	WORD-W	999	WIRING
CONCAT	0	@	999	WIRING
GFREAD	0	GBIT-R	999	WIRING
GFWRITE	0	GBIT-W	999	WIRING
GAREAD	−1	GWORD-R	999	WIRING
GAWRITE	−1	GWORD-W	999	WIRING
INC	1	INC	999	INC
DEC	1	DEC	999	DEC
CLEAR	1	CLEAR	999	CLEAR
MUX	0	MUX	999	MUX
DEMUX	0	DEMUX	999	DEMUX
BUS	0	BUS	999	BUS
ALU	1	ALU	999	ALU

memory representation for these parameters. If the user does not supply values for these parameters in a file named *daa.l*, the values from Tables 11, 12, 13, and 14 are used. Delay values of −1 are a special case of clock synchronization discussed in Section 5.3.2. Because these values were suggested by designers of MOS microprocessors attempting to design fast processors, the use of hardware is unlimited. The working-memory list of this table is given in Appendix B.

Table 13. DEFAULT DB-OPERATOR — PART 3

Hardware Opcode	Delay	VT Opcode	Max	Group
SELECT	−1	SELECT	999	BRANCH
ENDSEL	0	ENDSEL	999	BRANCH
DIVERGE	0	DIVERGE	999	BRANCH
MERGE	0	MERGE	999	BRANCH
ENTER	−1	ENTER	999	CONTROL
CALL	−1	CALL	999	CONTROL
PSTART	−1	PSTART	999	CONTROL
LEAVE	−1	LEAVE	999	CONTROL
RESTART	−1	RESTART	999	CONTROL
RESUME	−1	RESUME	999	CONTROL
TERMIN	−1	TERMIN	999	CONTROL
NOOP	0	NO.OP	999	NOOP
UNDEF	0	UNDEF	999	UNDEFINED
UNPREDICTABLE	0	UNPREDICTABLE	999	UNPREDICTABLE
PARITY	1	PARITY	999	PARITY
IS.RUNNING	0	IS.RUNNING	999	IS.RUNNING
DELAY	0	DELAY	999	SYNC
WAIT	0	WAIT	999	SYNC
DWAIT	0	TIME.WAIT	999	SYNC
STOP	0	STOP	999	SYNC

Table 14. DEFAULT FOLD AND DELAY

Fold Type	Proximity	Cost
ARITHMETIC	0.5	0.5
LOGICAL	0.5	0.75
REGISTER	0.5	0.5
Max Delay Per Control Step	50	

4.3 Technology-Independent Hardware Network

The VT representation and the constraints are used by the rules described in Chapter 5 to create a technology-independent, but technology-sensitive implementation description. In the CMU/DA system, the technology-independent description is given in a structure and control specification language, SCS[19] (soon to be replaced by DIF).[20]

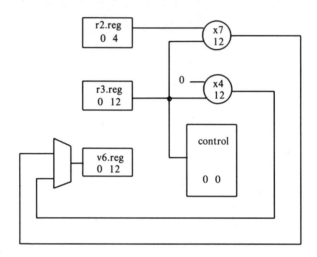

Figure 20. SAMPLE STRUCTURAL SPECIFICATION

The structural specification models the functional and logical structural levels as a network of modules. The modules communicate through links connected to module-interface ports. The ports may sink input, source output, or both sink input and source output. The control specification describes the control sequences that govern the behavior of the network.

Figure 20 shows a graphical SCS for the decoding loop of Figures 18 and 19. Each of the symbols represents a module. The circles are single function wiring modules to bring together or concatenate two sets of signals, the trapezoid is a multiplexer that gates one of its two inputs to its output, the small rectangles are registers, the large rectangle is the controller, and each line represents a link between the modules. Where the links join with the modules, a port is defined. Because the initial ISPS was so simple, the final design is just some wiring, the specified architectural registers, a multiplexer and the controller.

Figure 21 shows the control specification for the structural specification in Figure 20. Each control step, *cstep*, has a list of ports to activate with references to the VT operators that cause the activations. Branches in the

Section 4.3   Technology-Independent Hardware Network

```
controller: controller0;
EADD {v6} :
  cstep [1:1] {
    select controller0 (controller0.source0) {
      case [0] {
        cstep [2:2] {
          mux v6.input.mux (v6.input.source1)
            {vt_references: v6.x5};
          gfwrite v6.x4 {vt_references: v6.x5};
          concat v6.x4 {vt_references: v6.x4};
        }
      }

      case [1] {
        cstep [3:3] {
          mux v6.input.mux (v6.input.source0)
            {vt_references: v6.x8};
          fwrite v6.register {vt_references: v6.x8};
          gfwrite v6.x7 {vt_references: v6.x8};
          concat v6.x7 {vt_references: v6.x7};
        }
      }
    } {vt_references: v6.x2};
  }

  cstep [3:3] {
    leave EADD {vt_references: v6.x9};
  }

end;
```

Figure 21. SAMPLE CONTROL SPECIFICATION

control specification are shown by *select* operations on controller inputs with *case* statements to show possible values and actions. Thus, this specification shows that if the controller0.source0 input is 0, then the bottom input to the multiplexer is turned on, writing into the global variable v6, using the output of the concatenate, v6.x4. If the controller0.source0 input is 1, then the top input to the multiplexer is turned on, writing into the global variable v6, using the output of the concatenate, v6.x7.

Table 15. HARDWARE

Attribute	Values
module	
id	id number
type	REGISTER, MEMORY, CONSTANT, OPERATOR
atype	TC, OC, SM, US
word-left	starting word
word-right	ending word
bit-left	starting bit
bit-right	ending bit
value	value for constants
opt-flag	optimization flag
port	
id	id number
number	source number 0 or 1 nil otherwise
type	INPUT, OUTPUT, BIDIRECTIONAL
bit-left	starting bit
bit-right	ending bit
module	id of module
link	
source-port	id of source
source-bit-left	bit number of source starting bit
source-bit-right	bit number of source ending bit
dest-port	id of destination
dest-bit-left	bit number of destination starting bit
dest-bit-right	bit number of destination ending bit

For each design task the DAA generates a working-memory representation of the SCS representation, shown in Table 15. The first column provides the name of the attribute, while the second column describes the possible values. The way this representation is created and manipulated is discussed in Chapter 5. The working-memory listing for the SCS representation is given in Appendix C.

4.4 Bookkeeping Information

There are several other miscellaneous bookkeeping working-memory elements used to guide the context of the problem solving, to count the allocated multiplexer and bus ports, to remember the state of control-step assignment, and to remember the minimum size for VT operators.

The most important is the *context* working-memory element, which directs the focus of the DAA to a particular task. For example:

(context ^status active ^operation make-link ^object v6.x4.output 11 0 v6.x4.p1 v6.input.source1 11 0 v6.x5.p1)

This is a request for all the rules that make links from one module to another to make the best connection between bits 11 and zero of ports v6.x4.output and v6.input.source1.

The multiplexer, bus, and controller modules require an additional working-memory element to keep track of the allocated number of input and output ports. These working-memory elements should be subsumed into the module working-memory element in future versions of the DAA. The control-step, micro-step, and trim working-memory elements are temporary places to store information during the assignment of control steps to operators and the trimming of excessive bits from add, sub, and multiply operators. The meaning of these working-memory elements is fully described in Sections 5.3.2 and 5.3.3, which overview the control assignment and operator trimming. Table 16 shows the working-memory representation of these structures.

Table 16. MISCELLANEOUS

Attribute	Values
context	
status	pending, active, next, or not-processed
operation	operation goal
object	vector: objects to operate on
mux-port-count	
module	module of mux
icnt	number of input ports
bus-port-count	
module	module of bus
icnt	number of input ports
ocnt	number of output ports
controller-port-count	
module	module of controller
icnt	number of input ports
control-step	
cs	current control step
micro-step	
ms	current micro-control step
trim	
id	outnode id to trim
bits	possible bits to trim to

4.5 Representation Summary

In summary, Chapter 4 has presented three representations that have spanned the algorithmic to technology-independent-hardware-network levels in the CMU/DA system. The representations are defined by the algorithmic-description language, VT, the constraints placed on the design, and the technology-independent-hardware-network description language,

SCS. Both the modularity and the granularity of the data representations have affected how rules are created for the system. Chapter 5 shows how these representations are used by the DAA to design VLSI systems. It also groups the knowledge together into sections to highlight the modularity of the knowledge in the DAA.

Chapter 5

DAA KNOWLEDGE

Chapter 4 has discussed the VT, constraint, and SCS representations used by the DAA. These representations combined with the knowledge extracted from expert VLSI designers, using the interview method described in Chapter 3, have been used to design many computers including the MCS6502 and the IBM System/370.

An important goal of the research has been to understand how VLSI designers choose computer implementations. The problem division they use, which is shown in Figure 22, is grouped into general service functions, global implementation allocation, VT allocation, SCS allocation, and global improvements to the implementation. These subtasks are implemented in the DAA and discussed by functional sections in this chapter. A summary of those subtasks, rule types, and rule usage for the SCF3 design is provided in Table 17. The chapter also discusses two sets of rules that have been rewritten as procedures for efficiency. They estimate partitioning and cost information used to provide high-level floor plan information to the DAA.

The functional subtasks in the DAA differ in numbers of rules, frequency of use, and type of knowledge. Table 17 shows that the service

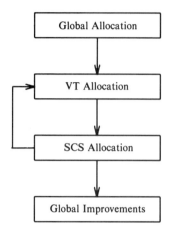

Figure 22. DAA SUBTASKS

functions and global allocation subtasks contain over half of the total rules and half the rules that extend the partial design. Furthermore, each of the remaining three subtasks contains about fifteen percent of the total rules and fifteen percent of the rules that extend the partial design. The most frequently used design rules are the port and link management rules, which follows logically because these are the rules that create the data path. These rules are discussed in the following section.

Table 17. RULES BY FUNCTION AND KNOWLEDGE TYPE

Section	Rules	Firings	Design	Context	Setup	Cleanup	Input	List	Test
Total DAA	314	8543	151	18	70	33	9	14	19
Service Functions	88	4901	70	7	0	3	0	4	4
Control management	12	1981	0	7	0	1	0	4	0
Module management	21	722	19	0	0	2	0	0	0
Port management	34	732	32	0	0	0	0	0	2
Link management	21	1466	19	0	0	0	0	0	2
Global Allocation	78	104	7	0	70	1	0	0	0
Memories	4	13	3	0	0	1	0	0	0
Registers	2	10	2	0	0	0	0	0	0
Constants	1	10	1	0	0	0	0	0	0
Controller	1	1	1	0	0	0	0	0	0
Technology database	70	70	0	0	70	0	0	0	0
Value Trace Allocation	55	1468	25	0	0	13	9	6	2
Initialization	(See Cleanup)								
Operator assignment	9	322	4	0	0	3	0	2	0
Operator trimming	13	64	0	0	0	2	9	0	2
Temporary registers	2	107	1	0	0	1	0	0	0
Register functions	6	18	6	0	0	0	0	0	0
ALU functions	25	957	14	0	0	7	0	4	0
SCS Allocation	47	1610	21	6	0	9	0	0	11
Register stability	2	173	4	0	0	1	0	0	3
Register allocating	9	26	4	2	0	0	0	0	2
Module stability	5	6	5	0	0	0	0	0	0
Module allocating	18	93	8	4	0	0	0	0	6
Cleanup	8	1312	0	0	0	8	0	0	0
Global Improvements	46	460	28	5	0	7	0	4	2
Unreferenced	6	58	5	0	0	1	0	0	0
Multiplexer cleanup	7	14	5	0	0	1	0	1	0
Multiple fan-out cleanup	4	12	0	3	0	1	0	0	0
Bus utilization	29	376	18	2	0	4	0	3	2
Final cleanup	(See Unreferenced and Multiplexer cleanup)								

5.1 Service Functions

An apprentice designer is first taught the tools to maintain or express a design. Along with simple bookkeeping, the designer is also taught to use the best or least expensive component. After this, various strategies of partitioning the problem into subtasks and managing the subtasks are acquired. These tools and skills, which are encompassed in the DAA as 88 rules to manage control, modules, ports, and links, are discussed in this section. An example of a rule to manage links is provided in Figure 23.

5.1.1 Control management. Like the designer dividing the task of design into subtasks, control management looks at ordering and selecting goals to complete a design. Control management involves two different types of rules. One creates goals to be later accomplished by the DAA; the other steps through outstanding goals, marks what has been done and focuses attention on what is left to do. Rules in the first category establish the major goals and subtasks described in this chapter. They recognize requests to do global allocation, VT allocation, SCS allocation, and global improvements. They also set up sub-goals to perform the tasks that are listed as subheadings in this chapter (See Figure 6). The rules used to step through the goals will either activate the next goal to be met, activate the next major task to be done, or decide no rule can perform a task (See Figure 7). Tasks that can not be performed are deleted, leaving an image of working memory in the current directory as *wm.?* for later debugging. Finally, there are rules to set up a certain goal context for every member of a list of working-memory elements. Table 17 shows that these rules compose about a fifth of the total rule firings, which follows logically from the observation that humans spend much time deciding what to do next.

5.1.2 Module management. The next type of management is module management. This set of rules has knowledge about creating, combining, and removing modules from the design. These rules allow the designer and thus other rules in the DAA, to think about creating, combining and removing modules as atomic units, rather than as individual parts. They also do the module bookkeeping discussed in the previous section. The create-module rules are activated by the goal *make-module*. If the module does not already exist, they create a module in working memory and write out a command to a collection of bookkeeping and service routines for the VT and SCS representations, VTDRIVE,[79] informing it that a module has been created. The move-module rules are activated by the goal *module-mv*. This goal is set when it is decided to make one module contain all the attributes and connections of another module, thus combining the modules. The attributes, ports, and links of one module are moved to another module, removing the first module. Lastly, the remove-module rules are activated by a goal to *module-rm*. This is done when a module is no longer needed in the design. Its links, ports, bookkeeping information, and finally the module itself are removed from the design.

5.1.3 Port management. The last two types of management, port and link, are closely coupled and shown in Table 17 to be heavily used. Port management deals with all features of ports, while link management deals with all features of links. They are closely related because many of the features of a port are defined by the link that connects to it. This section of rules encompasses much knowledge about making the least expensive connection from one port to another. Like the designer, they consider issues of existing data paths, expanding data paths, creating new data paths and multiplexers. A designer at INTEL said:

> "Reducing the number and length of connections is the
> single most important task in choosing an implementation."

Port management requires knowledge about creating new ports, increasing the size of old ports, creating multiplexers where needed, and rerouting connections through less expensive ports where possible. These rules are all activated by a request to *make-link*. If the source or destination port does not exist, one is created. If a second connection to an input port is requested, a multiplexer is created, or an existing one makes the connection. The only exception to this occurs when connections to the controller are requested; here just another input port is added to the controller. Wherever possible, idle existing ports are used. If a link is requested whose width exceeds the width of the port, the port and the module connected to the port are increased in width. If the expanded module is a multiplexer, then the module connected to the output of the multiplexer is also expanded. Lastly, whenever a request is made to link that could be satisfied by using an existing port or slightly expanding an existing port, the *make-link* request is modified to use this less expensive connection. Consider an example of this rerouting: a link from module A port A.out to multiplexer module B port B.in1 using bits four to one is requested, and there is already a link from A.out to module B port B.in0 using bits eight to two. The least expensive path is to modify the link request to use B.in0 by expanding its port by one bit.

Section 5.1   Service Functions   63

5.1.4 Link management. The last management function involves links and the expanding of existing links. This function works with the designer's knowledge codified in the previous section to make the least expensive connections between two ports.

IF:
    the most current active context is to create a link
    and no other link can be used to make the connection
THEN:
    create the link
    and mark the vt references for the source and destination ports

Figure 23. MAKE-LINK

These rules are also activated by the *make-link* goal showing the even closer coupling of this service with port management service. The link rules have knowledge about creating new links, using existing links, and expanding existing links. They are also responsible for indicating the control step when the link must be turned on to create a functioning design. They do this by informing VTDRIVE of the source and destination VT references[19] responsible for the creation of this link. Given that control steps can be assigned to links, then control steps can also be assigned to modules by looking at their input and output ports and the links connected to them. This information creates the control specification part of the SCS language described in Chapter 4. This also serves to create a history, which can be traced back to the algorithmic description, of why a module, port or link exists. The knowledge of the rules in this section involves looking at existing links between the source and destination ports and either making a new link, using an existing link, or expanding an existing link to fulfill the *make-link* goal.

5.2 Global Allocation

From the interviews in Chapter 3, the DAA was taught that global non-changing hardware is the first thing designers allocate. This partitioning makes the designer's job easier. The designers identified the base-variable storage elements, the database, the constraints, and the controller as global objects that probably would not change and should be allocated first. The base-variable storage elements are memories, registers declared globally across the design, formal parameters, and constraints. This section comprises 78 rules and is done under the context of *declared-variable-allocation*. Figure 24 shows this set of rules is fired only for the

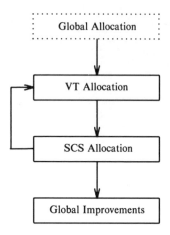

Figure 24. GLOBAL ALLOCATION

first VT-body allocated. It sets up all the modules and ports expected throughout the rest of the design and provides default constraints and timing information where none is supplied. The default constraints are given in Tables 11, 12, 13, and 14.

Figure 25 shows the design of the decoding loop of Figures 18 and 19 after global allocation has been done. Each of the symbols represents a module with the bit width given as the bottom pair of numbers. The small rectangles are registers, the small squares are constants, and the large rectangle is the controller. The registers r2.reg, r3.reg and v6.reg are allocated because of the ISPS definitions for cpage, i, and eadd, respectively. A controller and four constants, one for each of the constants shown in Figure 19, are also allocated. Although unseen in the figure, the technology database and constraints are initialized as given in Tables 11, 12, 13, and 14, and Appendix B.

This section overviews allocating memories, registers, constants, controller, and database by recognizing certain features in the VT representation. These rules, like all the other rules in this chapter, use the service function rules to do their bookkeeping. An example of a rule to allocate registers is provided in Figure 26.

Section 5.2   *Global Allocation*   65

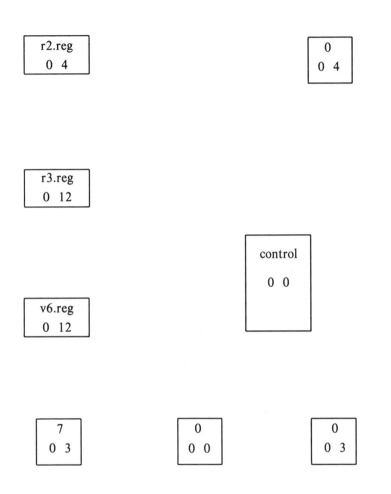

Figure 25. DESIGN AFTER GLOBAL ALLOCATION

5.2.1 Memories. Memory modules are allocated for VT-bodies that have word-left or word-right specified. For each memory module that is allocated, input, output, and address ports are also allocated.

5.2.2 Registers. Register modules are allocated for VT-bodies that do not have word-left or word-right attributes specified, but have bit-left or bit-right attributes. For each register module that is allocated, input and output ports are also allocated. If the parent VT-body of the VT-body causing a register to be allocated is not a section list, then the register

IF:
> the most current active context is declared variable allocation
> and there is a vtbody that is either a carrier, vtbody or section list
> and it is not an array
> and it has a non-zero width
> and the parent vtbody is a section list

THEN:
> create a register module
> and create an output port
> and create an input port

Figure 26. FIND-VTBODIES-FOR-REGISTERS

module is marked as a formal parameter and will later be removed. It will be removed because the convention used in the DAA forces all values into stable registers for the CALL and ENTER VT operations. Thus, two stable registers are unnecessarily connected.

5.2.3 Constants. Constant modules are allocated for outnodes with a constant type attribute. For each constant module only an output port is allocated.

5.2.4 Controller. Next a controller module is created for the design. The DAA currently has only rules that create single controller designs.

5.2.5 Technology database. Finally, any technology-sensitive database elements not read in from the *daa.l* file are created (See Figures 8 and 9). The default values are given in the section about the database and constraints.

5.3 Value Trace Allocation

Once the designer has allocated the global non-changing hardware, the next task is to partition the whole design into smaller blocks and select a partition for allocation. Within each partition, designers allocate clock phases, operators, registers, data paths and control logic. The design is partitioned by the CMU/DA system by using the natural boundaries of the VT-bodies. A VT-body can be chosen by the designer, or the closest unallocated VT-body to the current VT-body can be found by invoking the VTDRIVE command, *mknu*, as described in Section 5.5.1. The DAA allocates the clock phases, operators, registers, data paths and control logic in two subtasks, VT allocation and SCS allocation, which are shown in Figure 27. This allows the DAA to gather all the information about register usage in the VT allocation and then allocate registers and modules in the SCS allocation.

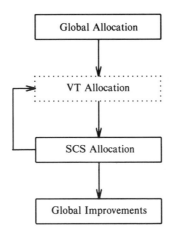

Figure 27. VT ALLOCATION

Figure 28 shows the design of the decoding loop of Figures 18 and 19 after VT allocation has been done. Each of the symbols represents a module. The circles are single function wiring modules to bring together or concatenate two sets of signals, the trapezoid is a multiplexer that gates one of its two inputs to its output, the small rectangles are registers, the small squares are constants, the large rectangle is the controller, and each of the lines represent a link between the modules. Where the links join with the modules, a port is defined. The state of the design shows the creation of the temporary registers, (x1.reg, x3.reg, x4.reg, x6.reg and x7.reg), the concatenation modules, (x7 and x4), a multiplexer, and many connections. Along with each connection the operator assignment to control steps and the VT references are specified, so that a control specification like Figure 21 can be generated. The temporary registers are created to latch values for testability reasons. The concatenation modules are created, rather than assigned as register attributes, because they are wiring instructions and cost nothing to create. The multiplexer is created because the output of either x7 or x4 needs to be stored in v6.reg. Finally a constant has been redrawn as an input to x4.

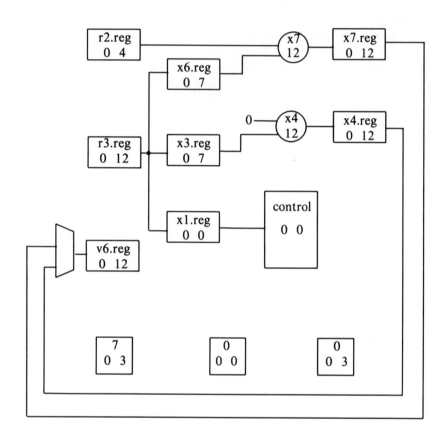

Figure 28. DESIGN AFTER VALUE TRACE ALLOCATION

The 55 VT allocation rules in this section assign operators to control phases, determine the minimum size needed to represent an operation, allocate temporary registers, and then associate VT operations to register modules or create ALU modules. This creation of register and ALU modules is not allocating hardware, but just mapping the VT representation into the uniform notation of the SCS representation. Rules described in the SCS section make the decision about allocating to hardware. An example of a rule to allocate temporary registers is provided in Figure 29.

5.3.1 Initialization. To reduce design-problem complexity, designers work with only one partition of a design and a listing of the currently allocated hardware. The DAA keeps an abbreviated form of the VT-bodies that are not currently being allocated in working memory, the full form of the VT-body being allocated, and the list of allocated hardware. This abbreviated form consists of only the *vtbody* working-memory structure described in Table 7. The DAA populates working memory with only the full form of the current VT-body and the input and output lists for the VT-bodies that are CALLED, ENTERED, LEFT, or RESTARTED from the current VT-body. To accomplish this there are rules to remove the abbreviated form of the VT-body and allow it to be replaced with the full form when necessary. These rules are activated by a goal *clean*, and will remove all VT-bodies not needed in working memory.

5.3.2 Operator assignment to control steps. After the unnecessary working-memory elements have been removed and the full VT-body is loaded into working memory, the next set of activated rules groups the VT operators into control steps. This is like the designer assigning clock steps for a maximally parallel design. If the control steps have not already been preallocated by VTDRIVE using the EMUCS[7] algorithm, the DAA will create the maximally parallel non-pipeline design. The rules step through the VT operators sequentially, summing the delays specified for them in the technology-sensitive database. When the sum becomes greater than the maximum allowed by the working-memory element *max-delay-per-control-step*, or the delay specified in the database is −1, a new full-control step is started. Stepping through the operators in order guarantees adherence to the data dependencies of the VT representation. The −1 delay value aligns memory, I/O, and changing context to a new VT-body operations to control-step boundaries. This synchronous and testable control structure allows LSSD techniques to be used by a layout program. Thus, these rules adhere to the maximum execution per control step, data dependency, and control dependency constraints necessary for a testable and correct design.

5.3.3 Operator trimming. The VT compiler takes a purist view of addition, subtraction and multiplication in the sense that arithmetic carries are always produced and then bit-read away when not needed. Now, this view may not seem bad, but if you want to add two 16-bit quantities and store the result in a 16-bit quantity you don't really need a 16-bit ALU with a carry. Even worse, if you have a chain of them, each will add bits to the width of the necessary ALUs. For example, 19-bit ALUs with carries could be created to add five 16-bit quantities and store them in a

16-bit quantity. DAA has two different types of rules for operator trimming. One type of rule looks for outputs to trim (See Figures 12 and 13), while the other trims inputs. These rules and the rules for temporary registers are activated by the goal of *temporary-variable-allocation*. Outputs of addition, subtraction, and multiplication operators are trimmed to the maximum size needed by all the places the output is fed. If an operator's output has been trimmed, the inputs of the operator are trimmed. The rules for trimming inputs are careful to trim only outnodes that are produced from other operations. Other types of outnodes are trimmed by creating a dummy outnode of the trimmed size and linking it to the trimmed operator. By trimming inputs that are the outputs of other operators, whole chains of operations are iteratively reduced to their minimum size.

5.3.4 Temporary registers. Designers at IBM stressed testable designs using a LSSD testing technique. To accomplish this it is necessary to place the results of all operations that cross clock boundaries into registers. The DAA allocates temporary-register modules for outnodes that are produced by VT operators regardless of control boundaries.

IF:
    the most current active context is temporary variable allocation
    and there is a produced value outnode
    and the outnode is not associated with an architectural register
THEN:
    create a temporary register module
    and create an output port
    and create an input port

Figure 29. FIND-OUTNODES-FOR-TEMPORARIES

Each register module is allocated input and output ports. Temporary-register allocation is activated by the same goal as operator trimming, *temporary-variable-allocation*, and fires after the produced outnodes are trimmed to the correct size. Register for operators in the same control cycle are removed by the register stability rules in Section 5.4.1.

5.3.5 Register functions. Next the designers allocate hardware operators, data paths and control logic. The DAA allocates operators either as register or ALU module attributes in the SCS representation. Both types are activated by the goal *operator-allocation*. In either case, the DAA will establish the data paths by the best port and link combination and mark

Section 5.3   Value Trace Allocation   71

the necessary VT references for the control logic.

This section discusses the rules particular to register modules, whereas the next section discusses ALU modules. A designer at AT&T Bell Laboratories pointed out that it is much less expensive to build a register that can increment/decrement itself than to build an ALU which adds/subtracts a register and the constant one and stores the result back in the same register (See Figure 4). Thus, there is a cost advantage to transforming operations that act on one register into register module attributes. This is also true for division, modulus, and multiplication by powers of two or shifting by a constant amount. As register attributes, these operations are wiring and control operations.

5.3.6 ALU functions. VT operations not allocated by the previous section are allocated as ALU module attributes. This allocation is not a real allocation to hardware, but a convenient mapping of the VT representation into the same notation as all the above hardware allocation. This makes it easier for the rules in the next section to combine these operations into ALUs. The types of operations handled are single source, dual source, branch, VT-body, controller, padding, and reading/writing registers/memories operations. The rules are activated by the same goal as in the previous section, *operator-allocation*, and constitute about a tenth of the total rule firings described in Table 17.

Addition, subtraction, division, modulus, multiplication, and shifting VT operators that are not allocated by the previous section are done in this section. The single and dual input operators addition, AND, concatenation, division, EQL, EQV, GEQ, GTR, LEQ, LSS, modulus, multiplication, NEQ, NOT, OR, SL0, SL1, SLD, SLI, SLR, SR0, SR1, SRD, SRI, SRR, subtraction, TST, XOR, and unary minus are transformed into goals of making an ALU module with the correct attribute and making links to the input and output ports. The modules, ports, and links are allocated by the rules discussed in the service function section of this chapter.

The DIVERGE and SELECT branch operators are transformed into goals of making a link from the control input to the controller and making links from each of the branch inputs to the corresponding DIVERGE or SELECT outputs.

The CALL, ENTER, and RESTART VT-body operators are transformed into goals of making links for each operator input to the corresponding input of the VT-body to be CALLED, ENTERED, or RESTARTED. This is half of the VT-body linking knowledge. The second half transforms the

LEAVE operator into goals of making links for each of its inputs to the corresponding output of the VT-body to be left. These two sets of rules have forced all values to be stable in registers over VT-body context switches. This aids in creating testable designs.

The DELAY, WAIT, and DWAIT controller operators are transformed into goals of making links to the controller.

The PAD0 and PADS operators are either transformed into goals of making a module with the correct attribute and making links to and from the input and output ports, or a goal of making a link from the input to the output register. This is required because in trimming operators to their minimum size, many pad operators become unnecessary.

The last group of rules to transform VT operators involves the reading and writing of parts of registers and memories. The BIT-R and GBIT-R operators are transformed into a goal of making a link from the input register with a given displacement to the output register. The BIT-W and GBIT-W operators are transformed into goals of making a link from the input register to the output register with a given displacement and making a link from the old-value register to the output register. The link from the old-value register supplies bits to the output register not otherwise supplied by the input register. The WORD-R and GWORD-R operators are transformed into goals of making a link from the word-offset register to the address port of the memory and the output port of the memory to the output register. The WORD-W and GWORD-W operators are transformed into goals of making links from the input register to the input port of the memory with a given displacement, from the old-value register to the input port of the memory, and from the word-offset register to the address port of the memory. The link from the old-value register supplies bits to the input port of the memory not otherwise supplied by the input register.

5.4 SCS Allocation

Once the designer has created control steps, operators, registers, data paths and control logic, the design is carefully examined for possible local improvements. Designers have identified removing stable registers, combining the remaining registers, removing modules that can be replaced by using the output of another module, and combining the remaining modules as possible local improvements. The knowledge in this section uses the VT representation, the constraints, and the SCS representation built by this and the previous VT allocation to make these decisions. Figure 30 shows the relationship of this tasks to the other tasks in the

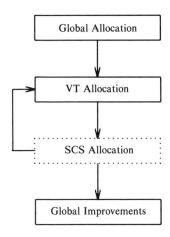

Figure 30. SCS ALLOCATION

DAA.

Figure 31 shows the design of the decoding loop of Figures 18 and 19 after SCS allocation has been done. The state of the design shows the removal of the temporary registers, x1.reg, x3.reg, x4.reg, x6.reg and x7.reg, and many connections. For this simple design, it turned out that all the registers are stable and not needed to keep the design testable. Thus, they were not made permanent. Also, the two wiring operations can not be combined because they do not have the same inputs. Had the operators not been wiring operators, they would have been examined for combination using the partition and cost estimators in Sections 5.5.1 and 5.5.2 and possibly combined.

Like the designers, the 47 rules in this section remove stable temporary registers, bind other registers to existing registers or allocate new registers, remove stable ALU modules, bind other ALU modules to existing ALUs or create new ALUs, and remove unnecessary working-memory elements. These rules are activated by the goal *fold-allocation*, except for the cleanup rules that are activated by the goal *cleanup*. An example of a rule to remove stable registers is provided in Figure 32.

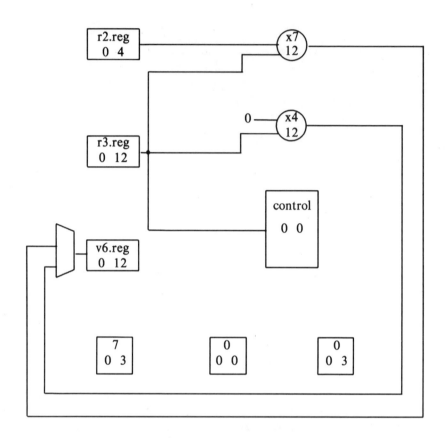

Figure 31. DESIGN AFTER SCS ALLOCATION

5.4.1 Register stability. Good hardware designers tend to create designs that are synchronous, that is, between control steps all values are latched into registers. However, a control step may be made up of several simple operations with each output feeding the next input in the chain. The DAA looks for temporary registers that may be in such a chain and removes them. To determine if a register is stable and can be removed, the DAA asks VTDRIVE if the control step for the input port of the register is the same as the control step for the output port. It also removes registers that are only fed from a constant, because these values are always available and do not need to be latched. Lastly, it removes registers that either have no outputs (See Figure 17) or whose only output feeds the controller. To

IF:
>the most current active context is fold allocation
>and there is a temporary register
>and there is a link to the temporary register
>and there is not a link from the temporary register

THEN:
>remove the temporary register
>and remove its link

Figure 32. FOLD-NO-OUTPUT-REGISTER

remove a register a goal is created to link from the input to the output of the register and to *module-rm* the register.

5.4.2 Register allocating. Designers are always looking to share registers efficiently. Efficient register sharing not only means finding a register of the correct size functionality and connectivity, but also one that will lay out well in later stages of the design synthesis. Estimators for cost and partitioning, which have been identified through the interviews in Chapter 3, are described in Sections 5.5.1 and 5.5.2. If a temporary-register module is not removed by the DAA in the previous section, then it must be combined with existing hardware or new hardware must be created. The DAA compares the temporary register with registers that have been bound to hardware modules by size, cost estimation, and proximity estimation (See Figure 16). It compares the temporary register to all bound registers of the same size and then to all bound registers of a larger size. Next, it compares the temporary register to all bound registers of a smaller size. If the cost estimation or the proximity estimation for the comparison is greater than the *cost* or *prox* attribute of the fold working-memory element for type REGISTER, then a goal of *module-mv* is created to combine the ports and attributes for the two register modules. Otherwise, the temporary register is bound to a new hardware register.

5.4.3 Module stability. Good hardware designers use the ability of a module to drive the output to several modules when appropriate. The DAA looks for temporary pad, concatenation, and single bit operators that are using the same inputs and attributes as other bound modules and removes the temporary ALU module. To remove a temporary module a *make-link* goal to move the output links from temporary module to the output of the bound module and a *module-rm* goal are created.

5.4.4 Module allocating. Designers are always looking to share ALUs efficiently. Efficient ALU sharing not only means finding an ALU of the correct size, functionality, and connectivity, but also one that will lay out well in later stages of the design synthesis. Estimators for cost and partitioning, which have been identified through the interviews in Chapter 3, are described in Sections 5.5.1 and 5.5.2. Designers will also combine ALUs by making tradeoffs between slower, more serial designs, and constraints on the number of modules. If a temporary ALU module is not removed by the rules in the previous section, then it must be combined with existing hardware or new hardware must be created. The DAA compares the temporary module with ALUs that have been bound to hardware modules by type, size, cost estimation, and proximity estimation (See Figure 5). It compares only modules of the same type, LOGICAL or ARITHMETIC. This is because an expert designer felt it best for testability not to combine LOGICAL operations into ALUs. Within each type it compares the temporary module to all bound modules of the same size. Then it compares the temporary module to all bound modules of a larger size. Next, it compares the temporary module to all bound modules of a smaller size. If the cost estimation or the proximity estimation for the comparison is greater than the *cost* or *prox* attribute of the fold working-memory element for type LOGICAL or ARITHMETIC, then a goal of *module-mv* is created to combine the ports and attributes for the two ALU modules. If module attributes can be allocated, the temporary module is bound to a new hardware ALU. Otherwise, the DAA will try to compare using less stringent *cost* and *prox* values. If no match is found, the DAA will serialize the operator by moving it to the next control step and adding one to all the control steps below it. If this fails to uncover a combination, the DAA will bind the module to hardware anyway and issue a warning message.

5.4.5 Cleanup. After all the hardware has been bound, the working memory is purged of all non-abbreviated VT-bodies. These rules, which are activated by a goal of *clean*, will remove all branches, control steps, lists, operators, outnodes (See Figure 10), trees, and VT-bodies not needed in working memory. This minimizes the number of working-memory elements. After the cleanup terminates, the process repeats with the VT allocation for the next VT-body until no more VT-bodies remain. The working-memory cleanup functions account for a sixth of the total rule firings (See Table 17); without them large designs could not be done within the virtual-memory address limitations of the VAX computer.

5.5 Estimators

As the DAA evolved from the prototype system, it became clear some things could not be handled well in rules. Primarily, these were actions that required summing, counting, or checking for set membership. Represented in OPS5 rules, these calculations were terribly inefficient; they could be much better represented as algorithms in a language like C. The decision making connected with the results of these estimators was kept in rules, but the calculation was moved into a C program. Through the interviews with designers in Chapter 3, these calculations evolved to focus attention on similarities of functions and connections in VT-bodies and modules. They mimic the *back of the envelope* floor plan on which designers base many of their decisions. These estimators have closely matched expert designer performance in partitioning and cost analysis experiments.

Primarily there are two estimators. One bridges the algorithmic to fabrication-dependent hardware-network level as a partitioner, while the other bridges the technology-independent to the technology-dependent hardware-network level as a cost function.

5.5.1 Partitioning. The first estimator is a partitioner that bridges the algorithmic to fabrication-dependent hardware-network levels. It is unlike most current partitioners of digital hardware[80, 81, 82, 83] because it does not require the specification of physical modules and their interconnections. That is, in our terms, it uses only the abstract and imprecise information contained in the ISPS description to make predictions of how a data path may be laid out. This layout guess is often referred to by designers as the *floor plan* of a chip. The floor plan partitioner attempts to share hardware effectively and minimize the interconnections between partitions.†

To share hardware operators effectively, the types of functions to be performed, the data path width, and the possibility of parallelism must be considered. To minimize interconnects, the common use of carriers and the direct passing of a value from one function to the next must be taken into account. In partitioning, all the above factors must be taken into account with the style of implementation and technology-sensitive information.

---

† This work was carried out under my supervision at AT&T Bell Laboratories in the summer of 1982 by Michael McFarland S. J.

They must be quantified, weighted, and combined to give a measure of what functions are similar and belong together, and what functions are not similar and can therefore be separated without great cost. A number of effective mathematical techniques have been developed to estimate similarities or distances between data points. One method for measuring the similarity of two objects[84] involves counting the number of properties they have in common. Suppose that two objects, 1 and 2, are to be compared. If A_1 is the set of properties belonging to object 1 and A_2 is the set of properties belonging to object 2, then an estimate of the similarity of the two objects can be based on the similarity of A_1 and A_2. A common measure of similarity is the so-called Jaccord, which is equal to

$$\frac{|A_1 \cap A_2|}{|A_1 \cup A_2|}$$

where for any set A, $|A|$ is the cardinality or the number of elements in A.

The measure of similarity developed here is analogous to the Jaccord concept described above. Instead of common properties, the algorithm counts the number of components and connections that could be shared by two functions, where components is an abstract category that can be made to correspond to gates or other hardware modules, depending on the implementation style. For each pair of functions in an ISPS description, two similarities are calculated. One, called the operator proximity, measures the commonality in the operations performed by the two functions. The other, called the register proximity, is a measure of the number of common register references and interconnections as a percentage of the total number of register references and interconnections used by the two functions. The complete similarity measure is then a weighted average of the operator proximity and register proximity. The way this average is formed and the weights used can be controlled by user-settable parameters.

The operator proximity between functions f_1 and f_2 is the cost of the sharable hardware divided by the cost of all the hardware needed to do all the computations in f_1 and f_2 separately.

The cost of a function f is the number of components needed to implement all the computational operators, such as adds, shifts, comparisons, and so on, used in the body of the function. The cost is based on the individual operators and the extent to which they can share the same hardware. For each type of operator i, there is a user-defined technology-sensitive cost c_i, which corresponds to the number of gates or components needed to implement the function, or some other measure of cost. For each pair of operator types, i and j, there is an overlap cost o_{ij}, which represents the amount of hardware that could be shared between the two operators. For example, if an adder takes six gates per bit and a subtractor eight gates, six of which can be shared with an adder, $c_{add} = 6$, $c_{sub} = 8$, and $o_{add,sub} = 6$.

For a set of operators S, the function cost(S) processes the operators in S one by one, finding the cost of adding the hardware capability for that operator to the hardware for the operators already processed. The cost of adding an operator to a given set is found by computing the cost of combining the operator with each operator in the set and then taking the minimum of those costs. The cost of combining operators i and j is $c_i + c_j - o_{ij}$.

The operator proximity between function f_1, with operators S_1, and f_2 with operators S_2 is then

$$\frac{\text{cost }(S_1) + \text{cost }(S_2) - \text{cost }(S_1 \cup S_2)}{\text{cost }(S_1 \cup S_2)}$$

because for any sets S and T, $S \cap T = S + T - S \cup T$. If f_1 and f_2 can be carried out simultaneously, then the operator proximity is lowered by a technology-sensitive factor that is set by the user.

The register proximity between functions f_1 and f_2 is the number of bits of carrier referenced in common, and of common interconnections, divided by the sum of the bits of carrier references and interconnections used in either one. Carriers in the ISPS description are weighted differently depending on what they stand for and how they are used. It is assumed that globals and arrays will remain as they are in the final design, so they are given full weight. Registers in which a value is directly passed between functions are also assigned full weight, because there must be some interconnection, whether the value is in that particular register or not. Non-architectural register references are downgraded by a technology-sensitive factor set by the user.

The total proximity is a weighted average of the operator proximity and the register proximity. This weighting can be done in two different ways. Static weighting simply uses two user-settable technology-sensitive weighting factors. Dynamic weighting takes into account the relative importance of operators and registers in the individual functions. Thus, if the functions are computation intensive, the operator proximity would weigh more heavily, whereas if they simply involve moving values around, the register proximity would weigh more heavily.

The partitioner has closely matched the performance of five expert designers participating in a partitioning experiment for the MCS6502 description.[85] The information is used by the DAA as a high-level floor plan to aid in making decisions about whether to create new modules or upgrade existing modules to contain the required connections and functionality. Further information is given in the discussion of register and module allocating in Sections 5.4.2 and 5.4.4.

5.5.2 Cost estimator. The second estimator is a hardware pricer that bridges the technology-independent to technology-dependent hardware-network levels. It is unlike the partition estimator because it requires that the system be specified by technology-independent modules and their interconnections. It provides information about what percentage of the required hardware for a new module already exists in another module. That is, in our terms, it uses only the abstract and imprecise information contained in the SCS description to make predictions of how much it would cost to upgrade the functions and interconnects of an existing module to contain a new module. Whereas the partition estimator gives a high-level floor plan, this estimator gives a much more local view, thus augmenting the high-level floor plan with more detailed information.

To price hardware modules, the types of functions to be performed, the data path width and the possibility of parallelism must be taken into account, along with the input and output data path connections. In pricing, all the above factors together with the style of implementation and technology-sensitive information must be considered. They must be quantified, weighted, and combined to give a measure of how expensive a new module would be as opposed to the cost of sharing an already existing module. One method for measuring the cost of adding a new object to an old object involves calculating the percentage of the new object's properties that already exists in the old object. Suppose that two objects, numbered 1 and 2, are to be compared. If A_1 is the set of properties belonging to object 1 and A_2 is the set of properties belonging to object 2, then an estimate of the increased cost of object 2, if it is to include object 1, can be based on the intersection of A_1 and A_2 scaled by A_1. This cost function is equal to

$$\frac{|A_1 \cap A_2|}{|A_1|}$$

The cost estimator developed here is analogous to the concept described above. For each pair of modules in a SCS description, two costs are calculated. One, called the module cost, measures the percentage of operations performed by the module 1 not in module 2. The other, called the port cost, measures the percentage of interconnections used by module 1 not in module 2. The complete cost measure is then a weighted average of the module cost and port cost.

The module cost between modules m_1 and m_2 is defined to be the cost of the hardware that can be used to do computations in both m_1 and m_2, that is, the sharable hardware, divided by the cost of the hardware needed to do the computations in m_1.

The cost of a module m is defined in the same way as the cost of a function f is defined in the previous section. The module cost between modules m_1, with operators M_1, and m_2 with operators M_2 is then

$$1 - \frac{\text{cost } (M_1 - M_2)}{\text{cost } (M_1)}$$

because for any sets S and T, $S \cap T = S - (S - T)$, where $S - T$ is the relative set difference defined by

$$A - B = \left\{ x \mid x \in A, x \neg\in B \right\}$$

If m_1 and m_2 are used in parallel, then the operator cost is -1.

The port cost between modules m_1 and m_2 is the cost of the hardware that can be used to do connections in both m_1 and m_2, that is, the sharable hardware, divided by the cost of the hardware needed to do the connections in m_1. The cost of a module m is the number of components needed to implement all the connections used in the module. The cost is based on the individual ports and the extent to which they can share the same hardware.

For a set of ports P, the connection cost(P) processes the ports in P one by one, finding the cost of adding the hardware capability for that port to the hardware for the ports already processed. When port references are made to wiring modules such as MUX, DEMUX, BUS, CONCAT, PAD0, or PADS, the ports of these modules are recursively taken into account.

The module cost between modules m_1, with ports P_1, and m_2 with ports P_2 is then

$$1 - \frac{\text{cost } (P_1 - P_2)}{\text{cost } (P_1)}$$

The total proximity is a weighted average of the module cost and the port cost. This weighting is done dynamically, taking into account the relative importance of functions and ports in the individual modules. It is given by:

$$\frac{\text{cost } (M_1) - \text{cost } (M_1 - M_2) + \text{cost } (P_1) - \text{cost } (P_1 - P_2)}{\text{cost } (M_1) + \text{cost } (P_1)}$$

Thus, if the modules are computation intensive, the module cost would weigh more heavily, whereas if they simply involve connections, the port cost would weigh more heavily.

The pricer has closely matched the performance of nine expert designers participating in a pricing experiment for the MCS6502 description.[86] The information is used by the DAA as a high-level floor plan to aid in making decisions about creating new modules or upgrading existing modules to contain the required connections and functionality. Further information is given in the discussion of register and module allocating in Sections 5.4.2 and 5.4.4.

5.6 Global Improvements

As a design nears completion, the designers start examining it for things that are no longer needed or could be better shared. This cleanup includes unused modules and ports, stable registers, duplicated multiplexer input ports, and modules that could be eliminated through use of common outputs. Finally, the designers examine the design for use of buses in high traffic areas. This set of 46 rules, like the global allocation set of rules described above and shown in Figure 33, is activated only once per design. Once these rules have fired, the design is done and no more VT-bodies can be processed without incurring errors.

Figure 34 shows the design of the decoding loop of Figures 18 and 19 after global improvements have been made. The state of the design shows

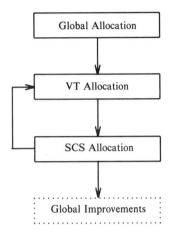

Figure 33. GLOBAL IMPROVEMENTS

the removal of the unused constants and the input ports to r2.reg and r3.reg. Because the initial ISPS was so simple, no multiplexer cleanup or bus allocation is required.

The DAA's global improvements remove unreferenced modules, remove unneeded registers, reduce multiplexer trees, use fan-out from modules, and allocate bus structures. An example of a rule to allocate bus structure is provided in Figure 35.

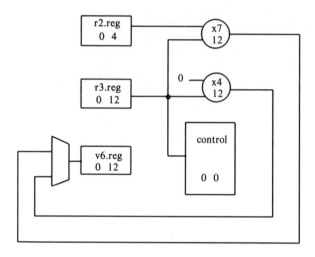

Figure 34. DESIGN AFTER GLOBAL IMPROVEMENTS

5.6.1 Unreferenced. Unnecessary modules and ports are created during the design process because some constants defined in the VT representation are not really needed in the hardware design; they only serve as offsets to wiring operations. Also the initial ISPS description may define carriers that are not really used. The DAA also tends to create a few extra multiplexer and bus ports in the act of finding the least expensive routing port. The first step towards cleaning up the design is the removal of unreferenced ports, modules, and formal parameter registers. These rules are activated by the goal *unreferenced*. If a port does not have a link connected to it, it is removed. If all the ports of a module are removed, the module is removed. If the module is a BUS or MUX type module, its bookkeeping information is also removed. Finally, formal-parameter register modules are removed, moving all their output links to the multiplexer or register that is feeding it. This is done because, as explained above, all values are stable on entry to VT-bodies. Thus, these registers are stable and are not needed to latch the input values.

5.6.2 Multiplexer cleanup. The removal of the formal-parameter registers tends to create trees of multiplexers, sometimes with duplicated input ports. The DAA has rules for the flattening out of the multiplexer tree structures and the removal of the duplicated input ports. It also removes multiplexers without output links and replaces multiplexers or buses with only one input port with links. These rules are active for the same goal as above, *unreferenced*.

5.6.3 Multiple fan-out cleanup. The DAA takes one more look for pad, concatenation, logical, and single bit modules that are using the same inputs and attributes as other bound modules to replace them with the ability of a module to drive output to several modules. It has to do this a second time, because the unreferenced and multiplexer cleanup can create more fan-out sharing.

5.6.4 Bus utilization. Now that everything has settled down, the DAA examines the design for possible bus usage.

IF:
    the most current active context is bus allocation
    and there is a module that is a multiplexer
    and there is another module that is also a multiplexer
    and there is a link from a non-bus module to the first multiplexer
    and there is a link from that module to the second multiplexer
    and there is a link from another non-bus module to the
        first multiplexer
    and there is a link from that module to the second multiplexer
THEN:
    place these connections on an idle bus

Figure 35. CONVERT-MUX-INPUTS-TO-BUS

Because buses are expensive in area and in power, they are used sparingly in VLSI design. Thus, the DAA tends to use buses only if there is a cost gain over using discrete multiplexers (See Figure 3). There are three types of bus rules, all activated by the goal *bus-allocation*. There are rules to propose that multiplexer inputs be moved to a bus, there are rules to find the best bus to place inputs on, and finally there are rules to add other inputs and outputs to a bus. Figure 36 shows an example of a multiplexer configuration on the left that would be allocated as the bus configuration on the right after these rules fire. Moving multiplexer inputs to a bus is proposed if there are two multiplexers, and on each multiplexer there are two identical inputs. Once this proposition has been made, buses containing both inputs are checked first, followed by buses containing only one input. The input ports of the bus and the proposed input ports are checked for timing collisions. If the bus is not idle, then this is followed by buses of the same, larger and smaller sizes. If none of the buses were idle

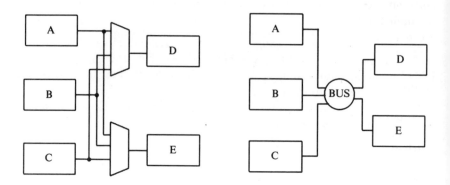

Figure 36. AN EXAMPLE BUS ALLOCATION

and the limit for the number of buses has not been exceeded, then a new bus is allocated. Otherwise whenever an idle bus is found, the inputs and outputs are optimally linked onto the bus.

Finally, there are three configurations of multiplexers, buses and modules that will be tested for moving to bus ports. If there is a module with an output to a multiplexer and a bus, and the bus has an output to the same multiplexer, then a check is made to see if the output from the module to the multiplexer can be moved onto the bus. If there is a module with an output to a multiplexer and a bus with an output to the same multiplexer, a check is made to see if the output from the module to the multiplexer can be moved onto the bus. Lastly, if there is a module with an output to a multiplexer and a bus, a check is made to see if the output from the module to the multiplexer can be moved onto the bus.

5.6.5 Unreferenced and final cleanup. After the multiple fan-out cleanup and bus allocation, the *unreferenced* goal is activated one last time for a final cleanup. See the above sections on unreferenced and multiplexer cleanup for more information.

5.7 Knowledge Summary

Chapter 5 has shown how designers partition the design task into four major tasks. These tasks involve global allocation of architectural modules and registers; local allocation of control steps, operators, registers, data paths and control; global allocation of operators and registers; and global improvements to the design. Within these tasks, a dominant goal is to partition the design into smaller, more manageable pieces. While making the design more manageable, it is also important to retain a global view of the hardware. Designers retain this global view by using high-level floor plans that are filled in as the design proceeds. This floor planning mechanism is composed of estimators for partitioning and pricing hardware. These estimators are concerned with connectivity and functionality of the networks of hardware. The designers feel it is important to design testable synchronous designs. This constraint is even more important than making the least expensive connections or optimally sharing hardware. After testability, the most important consideration is how the design will lay out or how to minimize connectivity. Thus, creating a good design is not just minimizing components, but paying careful attention to testability and connectivity.

The knowledge in the DAA system has been gathered using a KBES approach. The approach allowed the separation of expert knowledge from the reasoning mechanism. This facilitated the incremental addition of new rules and the refinement of old ones because the rules could be written to have minimal interaction with one another. The rules in the DAA system were also written with the goal of placing technology-sensitive constants in the working memory. Although these goals made it easier to add new rules and change the target technology, they did not take full advantage of the optimizations possible in the OPS5 system. The general design of the DAA system was not highly interactive because the Franz Lisp version of the OPS5 interpreter is slower than could be tolerated by the expert designers. If the DAA were to be rewritten, a more interactive environment could be attained by using either the BLISS version of OPS5, which is about 12 times faster than the Lisp version (.92 rules/second in the Lisp version versus 10.6 rules/second in the BLISS version for a major part of the MCS6502 design) or the OPS83 system. Because OPS5 is particularly bad at summing, counting, and checking set membership, to speed up the DAA the rules for partitioning and cost estimating were rewritten as C procedures. Two final burdens in using OPS5 were that the structure of working memory was *flat* and there was no facility for time profiling the execution of rules. A hierarchical working-memory representation would have decreased the use of variables as pointers in the rule memory and thus, decreased execution time. In writing rules, the order of the antecedents can vary the run time of the system by a factor of ten. A time profiler or the dynamic ability to change the internal representation of the rules in the Rete network would have corrected this problem. In summary, use of the KBES approach decreased the time it took to gather and verify the knowledge used by expert VLSI designers, but use of OPS5 increased the time it took to create designs.

Chapter 6

THE IBM SYSTEM/370 EXPERIMENT

After the DAA successfully designed a MCS6502 microprocessor, it had to be determined whether the system had also acquired knowledge about processor design in general. In this regard, an experiment was designed to see whether the DAA could design a processor substantially different and more complex than the MCS6502.

An ISPS description was chosen for the complete IBM System/370 from the descriptions maintained at Carnegie-Mellon University. This description included memory-management operations, channel controller I/O instructions, and all the 370 instructions, except the extended-precision floating point, the characters under mask, the edit and mark, and the packed-decimal instructions. The unmodified System/370 description, missing only a small percentage of the total 370, is more than 10 times larger than that of the MCS6502,† and it had not been used to build the

---

† The next bigger description, the Digital Equipment Corporation VAX 11/780, wouldn't compile through the VT compiler in a six megabyte address space

DAA. Important benefits of this choice are that a single-chip design of the 370 had been made at IBM, information is publically available, and Claud Davis, the design team manager and a key designer, was willing to critique the design. Thus, the experiment was a fair and convenient way to test the generality of the DAA's design knowledge.

This chapter presents an IBM System/370 design produced by DAA, D370, as well as a comparison with the IBM System/370 bipolar gate array micro-processor chip, μ370,[87] produced by Claud Davis. For each difference, possible changes in the CMU/DA system and the DAA are discussed.

6.1 The D370 Design

The experiment was begun by taking the ISPS description of the 370 (2,841 lines / 63,307 bytes) and translating it into VT notation (47,486 lines / 1,581,231 bytes), which comprised 6,078 operators (1,268 of which are branch operators), 154 constants, 251 VT-bodies, 51 map declarations, two section lists, 9,514 values (88 of which are declared registers), 2,692 outputs from VT-bodies, 14,240 input to VT-bodies (27 of which are formal inputs declared in the ISPS description), and 63,899 inputs to operators. The VT file was used as input to VTDRIVE, where high-level partition planning was done as discussed in Section 5.5.1. Then each VT-body was given to the DAA in turn. The DAA designed the D370, its version of the System/370, in 47 hours of CPU time on a VAX 11/780 with six megabytes of memory and two memory controllers. The D370 was designed without rule modifications or design iterations of any type.

The D370 is an IBM System/370 data-flow design using a $50x$ clock, where x is some scaled unit of time like μseconds, with multiplexer and bus style data paths. The DAA's constraints were set to produce a high-performance machine — that is, it could use as much hardware as required to allocate the data paths and retain maximum parallel operator usage. To meet this performance constraint, the D370 has eight-bit, 24-bit and 64-bit buses, 32-bit, 64-bit and 68-bit ALUs, a few discrete components, six memory arrays, and a great many architectural registers. This section discusses the functional blocks of the D370 design.

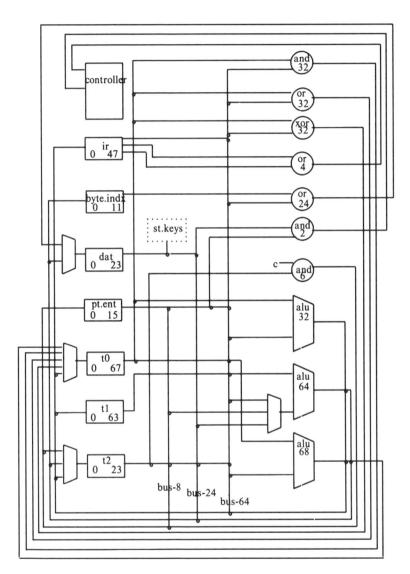

Figure 37. THE D370 DESIGN — PART 1

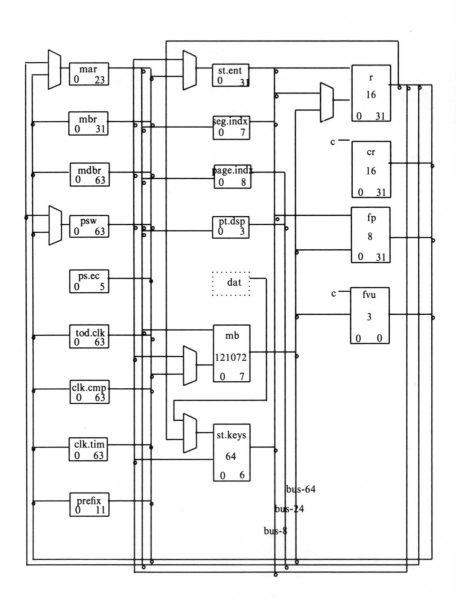

Figure 38. THE D370 DESIGN — PART 2

Section 6.1 The D370 Design 93

6.1.1 The three ALUs. The three ALUs and bus sizes arose from three different groups of data operations and major transfer widths in the IBM System/370. The basic busing style placed a temporary register before an input to each ALU and assumed the ALU latches its result so it can be read on the next clock phase transition. Thus, a two-phase clock set up the inputs to an ALU in one clock cycle and stored the result on the next clock cycle.

For clarity, the data path of the D370 is drawn in two separate figures, Figures 37 and 38. The connection between the two figures is through the eight-bit, 24-bit and 64-bit data buses. Figure 37 contains the arithmetic portion, the temporary registers, and the controller; Figure 38 shows the architectural registers, including the register arrays, such as the 16 general purpose registers *R*.

The 32-bit ALU is used for most of the arithmetic operations in the System/370 architecture. It can ADD, SUBTRACT, and COMPARE two binary numbers from the *T0* temporary register and the 64-bit bus. It gates a result out on this bus.

The 64-bit ALU is used for most of the address calculation operations and a few low-frequency operations, such as MULTIPLY and MODULUS in the System/370 architecture. It can ADD, SUBTRACT, COMPARE, MULTIPLY, MODULUS, and SHIFT RIGHT two binary numbers from the *T1* temporary register and a bus. It can gate a result out on either the 24-bit or 64-bit bus.

The 68-bit ALU is used for most of the floating-point operations in the System/370 architecture. This ALU can ADD, SUBTRACT, COMPARE, and SHIFT LEFT two binary numbers from the *T0* register and the 64-bit bus. Its result is gated onto the 64-bit bus.

6.1.2 The discrete components. Not all data manipulation is done in the ALUs. To aid debugging, one of out expert designers from INTEL keeps single-function logic outside the ALU. Thus, the logic instructions are implemented with separate distributed logic elements: the 32-bit AND, OR, and XOR.

Smaller logic elements are provided for a variety of functions. The four-bit OR takes input from two fields of the instruction register *IR* and feeds the result to the microcontroller. This aids in instruction decoding. The virtual storage system uses three discrete components. The 24-bit OR takes input from the byte index *BYTE.INDX* register and the 64-bit bus and places the result in the dynamic address translation *DAT* register. The two-bit AND takes input from the 24-bit bus and the page table entries *PT.ENT* and feeds the result to the microcontroller. The six-bit AND takes input from a group of constants and *T2* and places the result on the eight-bit bus.

6.1.3 The memory arrays. The D370 architecturally defines the primary memory *MB*, the storage keys *ST.KEYS*, the general purpose registers *R*, the control registers *CR*, the floating point registers *FP*, and the floating point error registers *FVU*.

Table 18. MEMORY ARRAYS IN THE D370 DESIGN

Abbreviation	Bits	Words	Address	Inputs	Outputs
MB	8	121072	24bb	8bb, 64bb	64bb
ST.KEYS	7	64	R, DAT	8bb	8bb
R	32	16	8bb	8bb, 64bb	ST.KEYS, 8bb, 24bb, 64bb
CR	32	16	Constants		64bb
FP	32	8	8bb	64bb	64bb
FVU	1	3	Constants	64bb	64bb

Table 18 lists the bit width of each memory array, the number of words in the array, and what buses *bb* or registers connect to the address, input, and output ports. Thus, there are 64 storage keys, each seven bits wide, with their address port connected to the general purpose registers and the dynamic address translation register. The input and output ports are connected to the eight-bit bus.

6.1.4 The architectural and temporary registers. The D370 architecturally defines many registers. Table 19 lists the bit width of each register and tells what buses or registers connect to the input and output ports. Thus, the instruction register is 48 bits wide with its input connected to the 64-bit bus and its output connected to the 64-bit bus and the four-bit OR described above.

Table 19. REGISTERS IN THE D370 DESIGN

Abbreviation	Name	Bits	Inputs	Outputs
MAR	memory address	24	24bb, 64bb	24bb, 64bb
MBR	memory buffer	32	64bb	8bb, 64bb
MDBR	memory double buffer	64	64bb	8bb, 64bb
PSW	processor status word	64	24bb, 64bb	8bb, 24bb, 64bb
PS.EC	extended code	6		64bb
TOD.CLK	clock	64	64bb	64bb
CLK.CMP	clock comparator	64	64bb	64bb
CPU.TIM	CPU timer	64	64bb	64bb
PREFIX	prefix	12	64bb	64bb
IR	instruction register	48	64bb	64bb, 4-bit OR
ST.ENT	segment table entry	32	8bb, 64bb	8bb
PT.ENT	page table entry	16	8bb	8bb, 64bb, 2-bit AND
SEG.INDX	segment index	8	24bb	8bb
PAGE.INDX	page index	9	24bb	24bb
PT.DSP	page table displacement	4	24bb	24bb
BYTE.INDX	byte index	12	24bb	24-bit OR
DAT	dynamic address translation	24	24bb, 24-bit OR	24bb, ST.KEYS
T0	temporary 0	64	64bb, 32-bit AND, OR, XOR	64bb, 32-bit ALU, 68-bit ALU, 32-bit AND, OR, XOR
T1	temporary 1	64	64bb	64bb, 64-bit ALU
T2	temporary 2	24	8bb, 24bb, 64bb	8bb, 24bb, 64bb, 6-bit AND

6.1.5 The control specification. A symbolic microcode word controls the D370. A microcode word is required for each cycle of the machine. The generation of either a PLA or ROM based micro-engine is possible in later phases of the design synthesis task. A sample sequence from the dynamic address translation VT-body has the 24-bit bus gating the *MAR* to the *MB* address port, the 64-bit ALU adding the *MAR* from the 24-bit bus and the temporary register *T1* and gating the result on the 64-bit bus to the *MB* input port, with the six-bit AND anding the temporary register *T2* and a constant, gating the result on the eight-bit bus to a field in the *PT.ENT*. This illustrates the high degree of parallelism possible in the D370.

6.2 The µ370 Design

The µ370 is an IBM System/370 micro-processor data flow on a single bipolar gate-array masterslice chip. It uses a 100-nanosecond cycle clock and is capable of executing 200,000 instructions per second. The physical chip is 7x7 mm and dissipates 2.3 watts. The plan was to use no more than 5000 wired circuits, 3 watts of power, and 200 pins. To meet these size and power constraints, the problem was divided into on-chip and off-chip sections. This section discusses the functional blocks of the µ370.

6.2.1 The on-chip functional block. The on-chip functional block has an eight-bit ALU, a 24-bit incrementer/decrementer, I/D, a 24-bit shifter, two nine-bit parity generators, 17 eight-bit working registers, two eight-bit buffer registers, a 16-bit status register, a 24-bit register, and hardware to calculate the next microcode address. These components are wired together with two fan-in eight-bit buses, a fan-out eight-bit bus, a bidirectional 16-bit bus, a fan-in 24-bit bus and a fan-out 24-bit bus. These are shown in Figure 39.

The eight-bit ALU can ADD, SUBTRACT, OR, AND, and XOR either binary or packed-decimal numbers. The arithmetic operations of the ALU can be controlled directly from a microcode field or indirectly through a status bit located in register *S*. This indirect control feature allows sharing of microprogramming routines for the ADD and SUBTRACT operations. Two eight-bit buses feed two eight-bit buffer registers, *A* and *B*, which feed the ALU. These two registers can selectively gate groups of four bits that correspond to hex digits within the byte or pass the complete byte to the ALU. This gating is used for decimal operations. The *A* register can also pass its eight bits, rotated by four bits to reverse its two hex digits. This is used by the pack and unpack instruction. The output of the ALU is placed on an eight-bit bus gated to all the working registers.

The 24-bit I/D is a special purpose adder that can add a 24-bit binary number with the constants 0, 1, 2, 3, −1, −2, or −3. The constant input is directly controlled from a microcode field. This I/D is dedicated to address calculations. An address is gated from a set of three working registers onto the 24-bit bus feeding the memory address register *MAR*. Any value present in this register is gated to the memory address bus MAB, the input

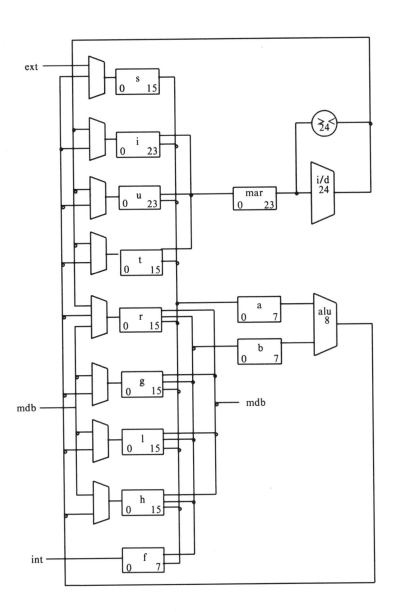

Figure 39. THE µ370 DESIGN

of the I/D, and a shifter (discussed below). The output of the I/D is placed on the 24-bit fan-out bus and gated back to the same three registers that fed the *MAR*.

The 24-bit shifter can do a few complex shift operations to produce a 16-bit result. This shifter is dedicated to handling the 12-bit address field used for page addressing in the virtual storage system. Input is gated from the *MAR*, and output is gated onto the 24-bit fan-out bus.

Two nine-bit parity generators check the parity of each data byte arriving at the chip and place it on the 16-bit memory data bus, MDB. They can also affix a parity bit to each data byte leaving the chip from the MDB.

The μ370 has 17 eight-bit registers, two eight-bit ALU buffer registers, a 16-bit status register, and a 24-bit *MAR*. The 17 eight-bit registers are grouped in functional pairs and triples. The *R*, *G*, *L*, and *H* register pairs are primarily the memory data registers, MDRs. Because the *G* register pair has special microcode branching capabilities, it is the OP code register. The *I* and *U* register triples are the program counter and operand register, respectively. The *T* register pair is the local-store address. The *I*, *U*, *T*, and *R* register groups can pass two bytes of data to one another with or without a displacement. This feature is part of the virtual storage management of the IBM System/370 architecture. The *S* register pair is a CPU status and serves as input to the microcontroller. *S1* can be set and reset by external inputs. The interrupt register *F* is also settable by external conditions.

A 54-bit microcode word controls the μ370. A microcode word is required for each cycle of the machine and is fetched during the last 75 nanoseconds of each cycle from the read only store, ROS. To select the next ROS word, a 16-bit address is generated in the first 25 nanoseconds of each cycle. Six bits of the ROS address are taken directly from the ROS word. The low order two bits are extracted from conditions within the chip. ROS fields dictate the internal conditions to be examined. The remaining bits are taken from the previous ROS address. However, if an external-trap bit is raised, the ROS address is forced to a specific value for a trap handler. Possible traps are parity errors, IPL request, page overflow, storage wrap, memory protect violation, stop request, and I/O control.

6.2.2 The off-chip functional block. The off-chip functional block has the architectural registers, two external memories, an I/O port, and the ROS. The μ370 chip uses 512 bytes of architectural registers. They are kept in a

Section 6.2 The μ370 Design 99

local store that can be accessed in 60 nanoseconds. The local store, which is limited to 64K bytes, is addressed by the *T0* and *T1* registers. The μ370 uses up to 16 megabytes of memory, which is addressed by the MAB, while data is gated on the MDB. *Read*, *write*, *memory-1*, *memory-2*, and *ready* signals allow up to two memories of any speed to be interfaced to the chip. If neither memory line is asserted, the MDB is connected to an I/O bus that uses the MAB to choose the I/O device being serviced.

6.3 The D370 and μ370 Design Comparison

The previous two sections have discussed the individual attributes of the two designs. This section brings those designs together by comparing and contrasting their differences. Claud Davis compared the two designs at IBM Poughkeepsie. During his career of over 25 years at IBM, Davis has worked on designs and managed teams of designers for the 701, 702, 7074 MA, 360/50, FAA, 360/67, and the μ370. His vast experience with the higher-performance processors and the μ370 made his critique valuable in two ways. First, we could determine what is needed for a single-chip IBM System/370 architecture; second, we could determine what is needed for a higher performance processor. Davis summarized his comparison thusly:[88]

> "The 370 data-flow we reviewed exhibited the quality I would expect from one of our better designers. The level of detail was what we call second level design. This encompasses all 'architected' registers, status latches and sufficient working registers to implement the functions defined by the instruction set. This level of design is independent of implementing technology.
>
> The review included a test for 'architected' registers, data path widths, latches for exceptional conditions, signs, and latches for temporary information in multi-cycle instructions.
>
> The assumptions for clocking and controls were examined and found to be consistent."

Relevant sections of the complete transcript[89] are included in Appendix D.

In the following discussion, differences are grouped by objectives and functional blocks including: ALUs, buses, memories, and registers, which are summarized in Table 20. For each difference possible changes in CMU/DA and DAA are discussed.

Table 20. IBM SYSTEM/370 — DESIGN DIFFERENCES

Design	D370	μ370
Objectives	High performance, technology sensitive, independent of power and I/O pins; paper design	Strict observance of technology criteria such as number of wired circuits, power, and I/O pins; working chips
ALUs	32-bit, 64-bit, and 68-bit; Binary numbers; hardware for virtual memory, floating point and multiply	Eight-bit and 24-bit; Binary and packed-decimal numbers; microcode for virtual memory, and multiply
Buses	Eight-bit, 24-bit, and 64-bit; bidirectional	Three 8-bit, a 16-bit, two 24-bit; fan-in, fan-out and bidirectional
Memories	12-byte buffer; single ported	Eight-byte buffer; single ported
Registers	Discrete	Memory array

6.3.1 The objectives. The objectives and testing of the two designs differed. The μ370's objective was to place a fully functional System/370 on a single chip, while observing such technology constraints as number of wired circuits, power, and I/O pins. The D370's objective was to design a high-performance System/370 sensitive to technology constraints, but independent of power and number of I/O pins. The μ370 was produced as a working chip, whereas the D370 is only a paper design.

6.3.2 The ALUs. The number, size, and type of functions supported by the ALUs in the two designs differed. The D370's design had one extra ALU that can be directly traced to the implementation of floating-point operations, while the μ370 planned a separate floating-point chip. In addition, the D370 implemented the dynamic address translation hardware, while the μ370 supplied a shifter that had a few complex shift patterns to aid in calculating the virtual address by using microcode.

The DAA does a high-level floor layout to help decide how to partition the algorithmic description, but this does not currently allow exclusion of functionality; the whole algorithmic description is implemented. Changes that modify the algorithmic description by including or excluding functionality are best made by changing the initial description or by having a postprocessor feed size constraints to the DAA.

The ALUs also differed in size. The μ370 serialized the 32-bit and 64-bit operations of the System/370 architecture into four or eight cycles through an eight-bit ALU. The DAA's constraints were set to design a high-performance processor, and thus the data paths were not serialized. Less than a dozen rules could be added to the DAA to allow it to serialize on ALU width. However, this change would be better made by adding a transformation to the CMU/DA system that removes a single-abstract data flow operation, and replaces it with several smaller ones.

Finally, the ALUs differed in the functions they provided. The μ370 has an ALU that can ADD and SUBTRACT packed-decimal numbers; the D370 performed these operations by adding hardware and microcode. The D370 has an ALU that can MULTIPLY, while the μ370 MULTIPLIED by SHIFTING and ADDING. Davis felt the choice of an ALU that could MULTIPLY was reasonable and consistent with the constraints used by the DAA.

6.3.3 The buses. The designs differed in the number, size and type of buses used. The μ370's eight-bit buses and 16-bit bus serve the same purpose as D370's eight-bit and 64-bit buses. The size difference is accounted for by the μ370 serializing the 32-bit and 64-bit operations of the System/370 architecture down to eight-bit operations, as discussed above. Also, the D370 uses bidirectional buses, where the μ370 used separate fan-in and fan-out buses. Davis felt the design choices made by the DAA were reasonable for the higher-performance D370 design, citing the IBM System/370 model 158 as an example of this style of busing.

6.3.4 The memories. The memory functional blocks differed only slightly. The D370 uses an eight-byte buffer with a four-byte memory data register, while the μ370 uses eight bytes of memory data registers. Davis felt this and even more elaborate cache schemes suited the higher-performance processors. He suggested the D370 use dual-ported memories for its general-purpose registers, to allow the use of two registers during the same cycle. Dual-ported memories would require a few rule changes, but would allow up to two memory array accesses during the same clock cycle. The rules for finding the address, input and output ports of memories would have to be enhanced to check for idle ports. All told, about 20 rules would have to be modified or extended to effect this change.

6.3.5 The registers. Both descriptions have about the same number of bytes of architectural and temporary registers. However, the μ370 groups all the architectural registers off chip in a fast local store, which can be thought of as memory. This would require a major change in the structure of the DAA. However, it could be accomplished simply, as a post-processor pass by the CMU/DA system as other technology-specific hardware is bound to the modules.

6.4 Summary of The System/370 Experiment

This chapter has shown the generality of design knowledge in the DAA by comparing and contrasting an IBM System/370 designed by an expert human designer, Claud Davis, against the design produced by the DAA. The differences were either explained and shown to be unimportant, or changes to the system were discussed. Davis felt the design produced by the DAA exhibited the quality he would expect from one of his better designers.

This chapter has shown the first large implementation design, automatically generated from an algorithmic description and constraints, that has been favorably critiqued by an expert designer. Furthermore, the design required 47 hours of CPU time, which with some work can be reduced by a factor of 12 to about 4 hours of CPU time. This clearly shows the dramatic improvement in CPU time for large designs obtained using methods that replace backtracking by match techniques. Finally, because this design was generated using synthesis techniques, it is possible to verify its operation by construction[90] and link it to the rest of the CMU/DA design environment.

Chapter 7
CONCLUSION

This thesis has shown how expert VLSI designers choose a MOS microcomputer's implementation and how a knowledge-based expert-system, the DAA, can mimic their results. It begins by showing that the KBES technique is a promising approach to design synthesis. Using a KBES approach has sped the development of the DAA by providing a framework that allows incremental addition of common-sense modular design knowledge and queries about the knowledge during the design task. This framework has increased the flexibility of the final system and reduced the execution time by replacing backtracking techniques with match techniques (the D370 design time can be reduced to about 4 hours of CPU time). Like a human designer, the DAA has become a better designer as its rule memory expands. The extraction, codification and testing of the expert designers' knowledge has facilitated a better understanding of VLSI design-synthesis, while providing another KBES system for computer scientists and knowledge engineers to examine. This KBES approach has opened the door to a whole new class of intelligent computer aided design tools.

The thesis has also examined how human designers are constantly partitioning the design task into more manageable subtasks. These tasks involve global allocation of architectural modules and registers, local allocation of control steps, operators, registers data paths and control, global allocation of operators and registers, and global improvements to the design. The designers retain a global view of the hardware by using high-level floor plans concerned with connectivity and functionality, that are updated as the design proceeds. More important than minimizing components, designers strive for testable synchronous designs that will lay out well. Thus, the thesis has shown that creating a good design is not just minimizing components, but paying careful attention to testability and connectivity.

The DAA has created efficient, testable and usable designs of many processors including the MCS6502 and the IBM System/370. The design of the MCS6502 was critiqued during the knowledge acquisition development phase of the DAA and thought to be a good design. A design of each of the small ISPS descriptions maintained at CMU showed that the system would produce a functionally correct design for a large number of test cases. The IBM System/370 was critiqued at IBM and exhibited the quality expected from one of IBM's better designers. This is the first research effort in implementation allocation that has been verified by expert designers as producing quality designs. We are no longer comparing designs by counting the number of components, but having experts comment on the quality of the design.

Finally, because the DAA is incorporated into the CMU/DA synthesis environment it is possible to verify a design's operation by construction, decrease the time it takes to design a chip, automatically provide multi-level documentation for the finished design, and create reliable and testable designs.

However, the research does not end here. The DAA system could be expanded and improved in a number of ways by adding more digital design knowledge or by improving the underlying expert system. New styles of allocation could be incorporated into the DAA such as pipeline, fault tolerant, or digital signal processing architectures. New transformations for minimizing bit widths could be added to the VTDRIVE and the DAA. New user interaction and constraint handling could be added to the DAA. New underlying expert systems could be used to increase the speed of designs (OPS83 is currently claimed to yield an increase in speed between 20 to one and 50 to one, which would open the door to an interactive DAA environment). New faster rule antecedents ordering could be achieved by dynamically changing the discrimination net at run time or providing the necessary measurement tools for rule performance.

Looking to the far horizon, a knowledge-acquisition system limited to implementation design could be developed, which allows expert designers to add and modify design rules and test examples for the DAA system. This would allow the personalization of the rule base for each designer. Looking further, the knowledge in the DAA could be used to develop a teaching environment for beginning designers. Finally, the DAA could be extended to start with a design, critique it, and help a designer modify it to meet constraints.

REFERENCES

[1] Mead, C. and Conway, L., *Introduction to VLSI systems,* Addison-Wesley Publishing Company, Reading, Massachussetts (1980).

[2] Director, S. W., Parker, A. C., Siewiorek, D. P., and Thomas, D. E., "A design methodology and computer aids for digital VLSI systems," *IEEE Transactions on Circuits and Systems* **cas-28**(7)(July, 1981).

[3] Thomas, D. E., Hitchock, C. Y. III, Kowalski, T. J., Rajan, J. V., and Walker, R., "Automatic Data Path Synthesis," *Computer* **16**(12) pp. 59-70 (December, 1983).

[4] Marwedel, P. and Zimmermann, G., *MIMOLA Software System User Manual,* 1, Institut Fur Informatik und Praktische Mathematik, Christian-Albrechts-Universitat Kiel (May, 1979).

[5] Hafer, L. J., *Automated data-memory synthesis: A format Method for the Specification, Analysis, and Design of Register-Transfer Level Digital Logic,* PhD thesis, Department of Electrical Engineering, Carnegie-Mellon University (June, 1981). Also in Design Research Center DRC-02-05-81

[6] Hafer, L., *Data — Memory Allocation in the Distributed Logic Design Style,* Masters thesis, Carnegie-Mellon University (December 21, 1977).

[7] Hitchcock, C. Y. III, "Automated Synthesis of Data Paths," CMUCAD-83-4, SRC-CMU Center for Computer-Aided Design, Carnegie-Mellon University (January, 1983).

[8] Tseng, C. J. and Siewiorek, D. P., "Facet: A Procedure for the Automated Synthesis of Digital Systems," *Proceedings of the Twentieth Design Automation Conference,* pp. 490-496 (June 27, 1983).

[9] Feigenbaum, E. A., *Knowledge Engineering: The Applied Side of Artificial Intelligence,* Computer Science Department, Stanford University (1980).

[10] Siewiorek, D. P., Bell, C. G., and Newell, A., *Computer Structures: Principles and Examples,* McGraw-Hill, New York (1982).

[11] Brooks, F. P. Jr., *The Mythical Man-Month,* Addison-Wesley, Reading, Mass. (1975).

REFERENCES

[12] Thomas, D. E., "The Automatic Synthesis of Digital Systems," *Proceedings of the IEEE* **69**(10) pp. 1200-1211 (October, 1981).

[13] Hafer, L. J., "Automated Synthesis of Digital Hardware," *IEEE Transactions on Computers* **C-31**(1) pp. 93-109 (January, 1982).

[14] Zimmerman, G., "The MIMOLA design system: a computer aided digital processor design method," *Proceedings of the Sixteenth Design Automation Conference*, pp. 65-72 ACM SIGDA and IEEE Computer Society DATC, (June 1979).

[15] Barbacci, M. R., Nagle, A. W., and Northcutt, J. D., *An ISPS Simulator*, Department of Computer Science, Carnegie-Mellon University (January 4, 1980).

[16] Barbacci, M. R., Barnes, G. E., Cattell, R. G., and Siewiorek, D. P., *The ISPS Computer Description Language*, Department of Computer Science, Carnegie-Mellon University (August 16, 1979).

[17] Snow, E. A., *Automation of Module Set Independent Register-Transfer Level Design*, PhD thesis, Department of Electrical Engineering, Carnegie-Mellon University (April 1978).

[18] McFarland, M. C., "The VT: A Database for Automated Digital Design," DRC-01-4-80, Design Research Center, Carnegie-Mellon University (December 1978).

[19] Vasantharajan, J., *Design and Implementation of a VT-Based Multi-level Representation*, Masters thesis, Department of Electrical Engineering, Carnegie-Mellon University (February 10, 1982).

[20] Bushnell, M., Geiger, D., Kim, J., LaPotin, D., Nassif, S., Nestor, J., Rajan, J., Strojwas, A., and Walker, H., "DIF: The CMU-DA Intermediate Form," CMUCAD-83-11, Department of Electrical and Computer Engineering, Carnegie-Mellon University (July, 1983).

[21] Wulf, W. A., Johnsson, R. K., Weinstock, C. B., Hobbs, S. O., and Geschke, C. M., *The Design of an Optimizing Compiler*, North Holland, New York (1977).

[22] Thomas, D. E., *The Design and Analysis of an Automated Design Style Selector*, PhD thesis, Department of Electrical Engineering, Carnegie-Mellon University (April 1977).

REFERENCES

[23] Sakallah, K. and Director, S. W., "An Activity-Directed Circut Simulation Algorithm," *Proceedings of the 1980 IEEE International Conference on Circuits and Computers*, pp. 1032-1035 IEEE, (October, 1980).

[24] Leive, G. W., *The Binding of Modules to Abstract Digital Hardware*, PhD Thesis Proposal, Department of Electrical Engineering, Carnegie-Mellon University

[25] Leive, G. W. and Thomas, D. E., *Module DataBase — User's guide*, Second Edition, Department of Electrical Engineering, Carnegie-Mellon University (October 1980).

[26] Thomas, D. E. and Leive, G. W., "Automating Technology Relative Logic Synthesis and Module Selection," *IEEE Transactions on Computer-Aided Design of Integrated Circuits and Systems*, pp. 94-105 IEEE, (April, 1983).

[27] Maly, W., Strojwas, A. J., and Director, S. W., "Fabrication Based Statistical Design of Monolithic IC's," *Proceedings of the International Symposium on Circuits and Systems*, (April, 1981).

[28] Forgy, C. L., *OPS5 User's Manual*, Department of Computer Science, Carnegie-Mellon University (July, 1981).

[29] Gillogly, J. J., "Performance Analysis of the Technology Chess Program," CMU-CS-78-109, Department of Computer Science, Carnegie-Mellon University (March 1978).

[30] Di Russo, R., *A Design Implementation Using the CMU-DA System*, Masters thesis, Department of Electrical Engineering, Carnegie-Mellon University (October 20, 1981).

[31] Parker, A. C., Thomas, D. E., Siewiorek, D. P., Barbacci, M. R., Leive, G., and Kim, J., "The CMU Design Automation System: an example of automated data path design," *Proceedings of the Sixteenth Design Automation Conference*, pp. 73-80 ACM SIGDA and IEEE Computer Society DATC, (June 1979).

[32] Aho, A. V., Johnson, S. C., and Ullman, J. D., "Code Generation for Expressions with Common Subexpressions," *Journal of the ACM* **21**(1) pp. 146-160 (January, 1977). Also in ACM Symposium on Principles of Programming Languages, pp 19-31, 1976.

REFERENCES

[33] McFarland, M. C., *Allocating Registers, Processors and Connections*. (September 5, 1981).

[34] Brown, H., Tong, C., and Foyster, G., "Palladio: An Exploratory Environment for Circuit Design," *Computer* **16**(12) pp. 41-56 (December, 1983).

[35] Newell, A. and Simon, H. A., "GPS, A Program that Simulates Human Thought," pp. 279-293 in *Computers and Thought*, ed. Feigenbaum, E. A. and Feldman, J. A., McGraw-Hill, New York (1963).

[36] Duda, R. O. and Gaschnig, J. G., "Knowledge-Based Expert Systems Come of Age," *Byte* **6**(9) pp. 238-281 (September, 1981).

[37] Feigenbaum, E. A., "The art of artificial intelligence: I. Themes and case studies of knowledge engineering," *Proceedings of the Fifth International Joint Conference on Artificial Intelligence*, pp. 1014-1029 Massachusetts Institute of Technology, (1977). Also in Expert Systems in the Microelectronic Age, Michie, D. (Ed.) Edinburgh: Edinburgh University Press, 1979, pp. 3-25. and in AFIPS Conference Proceedings, Vol. 47, pp. 227-240 (June, 1978)

[38] Rychener, M. D., "Knowledge-Based Expert Systems: A Brief Bibliography," CMU-CS-81-127, Department of Computer Science, Carnegie-Mellon University (June 26, 1981).

[39] McDermott, J. D., *Private Communication*. (March, 1984).

[40] Davis, R., "Interactive Transfer of Expertise: Acquisition of New Inference Rules," *Artificial Intelligence* **12** pp. 121-157 (August, 1979).

[41] Clancey, W. J., "Tutoring rules for guiding a case method dialogue," *International Journal of Man-Machine Studies* **11** pp. 25-49 (1979). Also in Proceedings of the Sixth International Joint Conference on Artificial Intelligence (1979), pp. 155-161

[42] Pople, H. E., Jr. and et al., "DIALOG: A Model of Diagnostic Logic for Internal Medicine," *Proceedings of the Fourth International Joint Conference on Artificial Intelligence*, pp. 848-855 (September, 1975).

[43] Shortliffe, E. H., *Computer Based Medical Consultations: MYCIN*, Elsevier, New York (1976).

[44] Clancey, W. J., Shortliffe, E. H., and Buchanan, B. G., "Intelligent Computer-Aided Instruction for Medical Diagnosis," *Proceedings of the 3rd Symposium on Computer Application in Medical Care*, pp. 175-183 (1979).

[45] Fagan, L., Kunz, J., Feigenbaum, E. A., and Osborn, J., "Representation of Dynamic Clinical Knowledge: Measurement Interpretation in the Intensive Care Unit," *Proceedings of the Sixth International Joint Conference on Artificial Intelligence*, pp. 260-262 (August 20-23, 1979).

[46] Bennett, J. S. and Engelmore, R. S., "SACON: A knowledge-based consultant for structural analysis," *Proceedings of the Sixth International Joint Conference on Artificial Intelligence*, pp. 47-49 Tokyo, (August 20-23, 1979).

[47] Duda, R. O., Gaschnig, J., and Hart, P. E., "Model design in the Prospector system for mineral exploration," pp. 153-167 in *Expert Systems in the Micro Electronic Age*, ed. Michie, D., Edinburgh University Press, Edinburgh (1979). Also in Pattern-Directed Inference Systems, Waterman, D. A. and Hayes-Roth, F. (Eds.) New York: Academic Press, 1978, pp. 203-222

[48] Feigenbaum, E A., Buchanan, B. G., and Lederberg, J., "On Generality and Problem Solving: A Case Study Using the DENDRAL Program," pp. 165-190 in *Machine Intelligence 6*, ed. Meltzer, B. and Michie, D., American Elsevier, New York (1971).

[49] Wipke, W. T., "Computer Planning of Research in Organic Chemistry," pp. 381-391 in *Computers in Chemical Education and Research*, ed. Ludena, E. V. Sabelli, N. H. and Wahl, A. C., Plenum Press, New York (1976).

[50] Gelernter, H. L. and et al., "Empirical Explorations of SYNCHEM," *Science*, pp. 1041-1049 (September 9, 1977).

[51] Grinberg, M. R., "A knowledge based design system for digital electronics," *Proceedings of the First Annual National Conference on Artificial Intelligence*, pp. 283-285 AAAI, (1980).

[52] Stallman, R. M. and Sussman, G. J., "Forward Reasoning and Dependency-Directed Backtracking in a System for Computer-Aided Circuit Analysis," *Artificial Intelligence* **9** pp. 135-196 (1977).

REFERENCES

[53] Brown, J., Burton, R., Bell, A. G., and Bobrow, R. J., "SOPHIE: A sophisticated instructional environment," AD-A010109 (December 1974). Distributed by NTIS (Dept. of Commerce)

[54] Stefik, M. J., "Inferring DNA Structures from Segmentation Data," *Artificial Intelligence* 11 pp. 85-114 North-Holland, (August, 1978).

[55] deKleer, J., "Qualitative and Quantitative reasoning in Classical Mechanics," pp. 11-30 in *Artificial Intelligence: An MIT Perspective*, ed. Winston, P. H. and Brown, R. H., Massachusetts Institute of Technology, Cambridge, Massachusetts (1979).

[56] Barstow, D. R., "An Experiment in Knowledge-Based Automatic Programming," *Artificial Intelligence* 12 pp. 7-1119 (August, 1979).

[57] Bennett, J. S. and Hollander, C. R., "Dart: An Expert System for Computer Fault Diagnosis," *Proceedings of the Seventh International Joint Conference on Artificial Intelligence*, pp. 843-845 (1981).

[58] McDermott, J., "R1: An Expert in the Computer Systems Domain," *Proceedings of the First Annual National Conference on Artificial Intelligence*, pp. 269-271 (1980).

[59] McDermott, J., "Domain Knowledge and the Design Process," *Proceedings of the 18th Design Automation Conference*, pp. 580-588 ACM IEEE, (June 29, 1981).

[60] Chisholm, I. H. and Sleeman, D. H., "An Aide for Theory Formulation," pp. 202-212 in *Expert Systems in the Micro Electronic Age*, ed. Michie, D., Edinburgh University Press, Scotland (1979).

[61] Erman, L. D. and Lesser, V. R., "HEARSAY-II: Tutorial Introduction and Retrospective View," CMU-CS-78-117, Department of Computer Science, Carnegie-Mellon University (May 1978).

[62] Nii, H. P. and Feigenbaum, E. A., "Rule-Based Understanding of Signals," pp. 483-501 in *Pattern-Directed Inference Systems*, ed. Waterman, D. A. and Hayes-Roth, F., Academic Press, New York (1978).

[63] van Melle, W., "A Domain Independent Production-Rule System for Consultation Programs," *Proceedings of the Sixth International Joint Conference on Artificial Intelligence*, pp. 923-925 (August 20-23, 1979).

[64] Weiss, S. M. and Kulikowski, C. A., "EXPERT: A System for Developing Consultation Models," *Proceedings of the Sixth International Joint Conference on Artificial Intelligence*, pp. 942-947 (August 20-23, 1979).

[65] Balzer, R., Erman, L. D., London, P., and Williams, C., "HEARSAY-III: A Domain-Independent Framework for Expert Systems," *Proceedings of the First Annual National Conference on Artificial Intelligence*, pp. 108-110 (1980).

[66] Reboh, R., "The Knowledge Acquisition System," Final Report, SRI Project 6415, Artificial Intelligence Center, SRI International, Menlo Park, CA (September, 1979). Also in A Computer-Based Consultant for Mineral Exploration, Duda, R. O., et al.

[67] Buchanan, B. G. and Feigenbaum, E. A., "DENDRAL and Meta-DENDRAL: their applications dimensions," *Artificial Intelligence* **11** pp. 5-24 (1978).

[68] Clancey, W. J. and Letsinger, R., "Neomycin: Reconfiguring a Rule Based Expert System for Application to Teaching," *Proceedings of the Seventh International Joint Conference on Artificial Intelligence*, pp. 829-835 (1981).

[69] deKleer, J., Doyle, J., Steele, G. L., and Sussman, G. J., "Explicit Control of Reasoning," pp. 93-116 in *Artificial Intelligence: An MIT Perspective*, ed. Winston, P. H. and Brown, R. H., Massachusetts Institute of Technology, Cambridge, Massachusetts (1979).

[70] Rychener, M. D., *OPS3 production system language tutorial and reference manual*, Department of Computer Science, Carnegie-Mellon University (1980).

[71] Waterman, D. A., "User-Oriented Systems for Capturing Expertise: A Rule-Based Approach," pp. 26-34 in *Expert Systems in the Microelectronic Age*, ed. Michie, D., Edinburgh University Press, Edinburgh (1979).

REFERENCES

[72] Ditzel, D., *Private Communication.* (July, 1981).

[73] Maul, M., *Private Communication.* (July, 1981).

[74] Williams, G., *Private Communication.* (July, 1981).

[75] Wilson, A., *Private Communication.* (July, 1981).

[76] Bourne, S. R., *UNIX Circuit Design System.* First Edition, Bell Telephone Laboratories, Incorporated, Murray Hill, New Jersey (December 1980).

[77] Newell, A., "Heuristic programming: ill-structured problems," pp. 360-414 in *Progress in Operations Research 3*, ed. Aronofsky, J., Wiley, New York (1969).

[78] Rose, M. A., *Structured Control Flow: An Architectural Technique for Improving Control Flow Performance,* Masters thesis, Department of Electrical Engineering, Carnegie-Mellon University (November, 1983).

[79] Gatenby, D. A., *Digital Design from an Abstract Algorithmic Representation: Design and Implementation of A Framework for Interactive Design,* Masters thesis, Department of Electrical Engineering, Carnegie-Mellon University (October 9, 1981).

[80] Kernighan, B. W. and Lin, S., "An Efficient Heuristic Procedure for Partitioning Graphs," *Bell Sys. Tech. J.* **49**(2) pp. 291-308 (1970).

[81] Schweikert, D. G. and Kernighan, B. W., "A Proper Model for the Partitioning of Electrical Circuits," *Proc. 9th Design Automation Workshop*, pp. 57-62 ACM IEEE, (1972).

[82] Breuer, M. A., "A Class of Min-Cut Placement Algorithms," *Proc. 13th Design Automation Workshop*, pp. 284-290 ACM IEEE, (1976).

[83] Payne, T. S. and vanCleemput, W. M., "Automated Partitioning of Hierarchically Specified Digital Systems," *Proc. 19th Design Automation Conf.*, pp. 182-192 ACM IEEE, (1982).

[84] Sokol, R. S. and Sneath, P. H. A., *Principles of Numerical Taxonomy,* W. H. Freeman and Co., San Francisco (1963).

[85] McFarland, M. C., "Computer-Aided Partitioning of Behavioral Hardware," *Proceedings of the Twentieth Design Automation Conference*, pp. 472-478 (June, 1983).

[86] Kowalski, T. J., *Computer-Aided Cost Estimation from Implementation Specifications.* (unpublished).

[87] Davis, C., Maley, G., Simmons, R., Stroller, H., Warren, R., and Wohr, T., "IBM System/370 Bipolar Gate Array Micro-Processor Chip," *Proceedings of the International Conference on Circuits and Computers* **2**(2) pp. 669-673 (October, 1980).

[88] Davis, C., *Personal letter to Dr. D. E. Thomas.* (August 12, 1983).

[89] Kowalski, T. J., *The VLSI Design Automation Assistant: The IBM 370 Critique,* Department of Electrical and Computer Engineering, Carnegie-Mellon University (September, 1983).

[90] McFarland, M. C., *Mathematical Models for Verification in a Design Automation System,* Phd thesis, Carnegie-Mellon University (July, 1981). Also in Design Research Center DRC-02-06-81

Appendix A

WORKING-MEMORY VT

The following working-memory listing shows the same decoding loop as the ISPS and VT descriptions in Figures 18 and 19. Parentheses are used to delimit the boundaries of working-memory elements. The type of the working-memory element is given by the first name after the opening parenthesis. It corresponds to the names following the keyword *literalize* in Tables 7, 8, and 9. The type is followed by its attribute-value pairs. By looking at the ISPS and VT descriptions you can see the direct translation of the ISPS carriers CPAGE and I into VT-bodies, constants and expression results into outnodes, and the procedure EADD into a VT-body with a list of operators and outnodes. This is the form in which the design task is given to DAA.

(vtbody ^id s1 ^type s ^isp-name EXAMPLE ^vt 1)
(vtbody ^id r2 ^type r ^parent s1 ^bit-left 0 ^bit-right 4 ^isp-name CPAGE ^vt 2)
(vtbody ^id r3 ^type r ^parent s1 ^bit-left 0 ^bit-right 11 ^isp-name I ^vt 3)
(outnode ^id *.c1 ^type c ^bits 4 ^value 7 ^vt *)
(outnode ^id *.c2 ^type c ^bits 5 ^value 0 ^vt *)
(outnode ^id *.c3 ^type c ^bits 1 ^value 0 ^vt *)
(outnode ^id *.c4 ^type c ^bits 4 ^value 0 ^vt *)
(vtbody ^id v6 ^type v ^parent s1 ^entity-flag global ^bit-left 0 ^bit-right 11 ^isp-name EADD ^inputs Iv6 ^operators OPv6 ^outputs Ov6 ^calls Cv6 ^opcalls OPCv6 ^attributes ATv6 ^control-steps-assigned t ^vt 6)
(lists ^id Iv6 ^list v6.i1 v6.i2 v6.i3 ^vt 6)
(outnode ^id v6.i1 ^type i ^carrier-id r3 ^bits 12 ^isp-name I ^vt 6)
(outnode ^id v6.i2 ^type i ^carrier-id v6 ^bits 12 ^isp-name EADD ^vt 6)
(outnode ^id v6.i3 ^type i ^carrier-id r2 ^bits 5 ^isp-name CPAGE ^vt 6)
(lists ^id OPv6 ^list v6.x1 v6.x2 v6.x9 ^vt 6)
(operator ^id v6.x1 ^opcode BIT-R ^inputs Iv6.x1 ^outputs Ov6.x1 ^control-step-begin 1 ^control-step-end 1 ^parent v6 ^vt 6)
(lists ^id Iv6.x1 ^list v6.i1 *.c1 ^vt 6)
(lists ^id Ov6.x1 ^list v6.x1.p1 ^vt 6)
(outnode ^id v6.x1.p1 ^type p ^bits 1 ^isp-name PB ^vt 6)
(operator ^id v6.x2 ^opcode SELECT ^inputs Iv6.x2 ^outputs Ov6.x2 ^control-step-begin 1 ^control-step-end 1 ^parent v6 ^branches Bv6.x2 ^vt 6)
(lists ^id Iv6.x2 ^list v6.x1.p1 ^vt 6)
(lists ^id Ov6.x2 ^list v6.x2.p1 ^vt 6)
(outnode ^id v6.x2.p1 ^type p ^carrier-id v6 ^bits 12 ^isp-name EADD ^vt 6)
(lists ^id Bv6.x2 ^list v6.x2.b1 v6.x2.b2 ^vt 6)
(branch ^id v6.x2.b1 ^type b ^parent v6.x2 ^activations ACv6.x2.b1 ^operators OPv6.x2.b1 ^inputs Iv6.x2.b1 ^vt 6)
(lists ^id ACv6.x2.b1 ^list 0:0 ^vt 6)
(lists ^id OPv6.x2.b1 ^list v6.x3 v6.x4 v6.x5 ^vt 6)
(operator ^id v6.x3 ^opcode BIT-R ^inputs Iv6.x3 ^outputs Ov6.x3 ^control-step-begin 2 ^control-step-end 2 ^parent v6 ^vt 6)
(lists ^id Iv6.x3 ^list v6.i1 *.c4 ^vt 6)
(lists ^id Ov6.x3 ^list v6.x3.p1 ^vt 6)
(outnode ^id v6.x3.p1 ^type p ^bits 7 ^isp-name PA ^vt 6)
(operator ^id v6.x4 ^opcode @ ^inputs Iv6.x4 ^outputs Ov6.x4 ^control-

step-begin 2 ^control-step-end 2 ^parent v6 ^vt 6)
(lists ^id Iv6.x4 ^list *.c2 v6.x3.p1 ^vt 6)
(lists ^id Ov6.x4 ^list v6.x4.p1 ^vt 6)
(outnode ^id v6.x4.p1 ^type p ^bits 12 ^vt 6)
(operator ^id v6.x5 ^opcode GBIT-W ^inputs Iv6.x5 ^outputs Ov6.x5
  ^control-step-begin 2 ^control-step-end 2 ^parent v6 ^vt 6)
(lists ^id Iv6.x5 ^list v6.i2 v6.x4.p1 *.c3 ^vt 6)
(lists ^id Ov6.x5 ^list v6.x5.p1 ^vt 6)
(outnode ^id v6.x5.p1 ^type p ^carrier-id v6 ^bits 12 ^isp-name EADD
  ^vt 6)
(lists ^id Iv6.x2.b1 ^list v6.x5.p1 ^vt 6)
(branch ^id v6.x2.b2 ^type b ^parent v6.x2 ^activations ACv6.x2.b2
  ^operators OPv6.x2.b2 ^inputs Iv6.x2.b2 ^vt 6)
(lists ^id ACv6.x2.b2 ^list 1:1 ^vt 6)
(lists ^id OPv6.x2.b2 ^list v6.x6 v6.x7 v6.x8 ^vt 6)
(operator ^id v6.x6 ^opcode BIT-R ^inputs Iv6.x6 ^outputs Ov6.x6
  ^control-step-begin 3 ^control-step-end 3 ^parent v6 ^vt 6)
(lists ^id Iv6.x6 ^list v6.i1 *.c4 ^vt 6)
(lists ^id Ov6.x6 ^list v6.x6.p1 ^vt 6)
(outnode ^id v6.x6.p1 ^type p ^bits 7 ^isp-name PA ^vt 6)
(operator ^id v6.x7 ^opcode @ ^inputs Iv6.x7 ^outputs Ov6.x7 ^control-
  step-begin 3 ^control-step-end 3 ^parent v6 ^vt 6)
(lists ^id Iv6.x7 ^list v6.i3 v6.x6.p1 ^vt 6)
(lists ^id Ov6.x7 ^list v6.x7.p1 ^vt 6)
(outnode ^id v6.x7.p1 ^type p ^bits 12 ^vt 6)
(operator ^id v6.x8 ^opcode GBIT-W ^inputs Iv6.x8 ^outputs Ov6.x8
  ^control-step-begin 3 ^control-step-end 3 ^parent v6 ^vt 6)
(lists ^id Iv6.x8 ^list v6.i2 v6.x7.p1 *.c3 ^vt 6)
(lists ^id Ov6.x8 ^list v6.x8.p1 ^vt 6)
(outnode ^id v6.x8.p1 ^type p ^carrier-id v6 ^bits 12 ^isp-name EADD
  ^vt 6)
(lists ^id Iv6.x2.b2 ^list v6.x8.p1 ^vt 6)
(operator ^id v6.x9 ^opcode LEAVE ^call v6 ^inputs Iv6.x9 ^control-
  step-begin 3 ^control-step-end 3 ^parent v6 ^vt 6)
(lists ^id Iv6.x9 ^list v6.x2.p1 ^vt 6)
(lists ^id Ov6 ^list v6.o1 ^vt 6)
(outnode ^id v6.o1 ^type o ^carrier-id v6 ^bits 12 ^isp-name EADD ^vt
  6)
(lists ^id Cv6 ^list v6 ^vt 6)
(lists ^id OPCv6 ^list v6.x9 ^vt 6)
(lists ^id ATv6 ^list tc !2 !EFFECTIVE.ADDRESS! ^vt 6)

Appendix B

WORKING-MEMORY USER PARAMETERS

The following working-memory listing shows the default technology database and user parameters. If the user does not supply values for these parameters in a file named *daa.l*, these values are used. Because these values were suggested by designers of MOS microprocessors attempting to design fast processors, the use of hardware is unlimited. Parentheses are used to delimit the boundaries of working-memory elements. The type of the working-memory element is given by the first name after the opening parenthesis. It corresponds to the names following the keyword *literalize* in Table 10. The type is followed by its attribute-value pairs. By looking at Tables 11, 12, 13, and 14 you can see the direct translation of the values into working-memory elements. This is the form in which the technology database and constraints are given to DAA.

(db-operator ^hw-opcode PLUS ^control-step-delay 1 ^vt-opcode +
 ^allocated 0 ^maximum 999 ^group ARITHMETIC)
(db-operator ^hw-opcode MINUS ^control-step-delay 1 ^vt-opcode -
 ^allocated 0 ^maximum 999 ^group ARITHMETIC)
(db-operator ^hw-opcode MULT ^control-step-delay 1 ^vt-opcode *
 ^allocated 0 ^maximum 999 ^group ARITHMETIC)
(db-operator ^hw-opcode DIV ^control-step-delay 1 ^vt-opcode /
 ^allocated 0 ^maximum 999 ^group ARITHMETIC)
(db-operator ^hw-opcode MOD ^control-step-delay 1 ^vt-opcode MOD
 ^allocated 0 ^maximum 999 ^group ARITHMETIC)
(db-operator ^hw-opcode UMINUS ^control-step-delay 1 ^vt-opcode --
 ^allocated 0 ^maximum 999 ^group ARITHMETIC)
(db-operator ^hw-opcode AND ^control-step-delay 1 ^vt-opcode AND
 ^allocated 0 ^maximum 999 ^group LOGICAL)
(db-operator ^hw-opcode EQV ^control-step-delay 1 ^vt-opcode EQV
 ^allocated 0 ^maximum 999 ^group LOGICAL)
(db-operator ^hw-opcode OR ^control-step-delay 1 ^vt-opcode OR
 ^allocated 0 ^maximum 999 ^group LOGICAL)
(db-operator ^hw-opcode XOR ^control-step-delay 1 ^vt-opcode XOR
 ^allocated 0 ^maximum 999 ^group LOGICAL)
(db-operator ^hw-opcode NOT ^control-step-delay 1 ^vt-opcode NOT
 ^allocated 0 ^maximum 999 ^group LOGICAL)
(db-operator ^hw-opcode EQL ^control-step-delay 1 ^vt-opcode EQL
 ^allocated 0 ^maximum 999 ^group RELATIONAL)
(db-operator ^hw-opcode NEQ ^control-step-delay 1 ^vt-opcode NEQ
 ^allocated 0 ^maximum 999 ^group RELATIONAL)
(db-operator ^hw-opcode LSS ^control-step-delay 1 ^vt-opcode LSS
 ^allocated 0 ^maximum 999 ^group RELATIONAL)
(db-operator ^hw-opcode LEQ ^control-step-delay 1 ^vt-opcode LEQ
 ^allocated 0 ^maximum 999 ^group RELATIONAL)
(db-operator ^hw-opcode GEQ ^control-step-delay 1 ^vt-opcode GEQ
 ^allocated 0 ^maximum 999 ^group RELATIONAL)
(db-operator ^hw-opcode GTR ^control-step-delay 1 ^vt-opcode GTR
 ^allocated 0 ^maximum 999 ^group RELATIONAL)
(db-operator ^hw-opcode TST ^control-step-delay 1 ^vt-opcode TST
 ^allocated 0 ^maximum 999 ^group RELATIONAL)
(db-operator ^hw-opcode SR0 ^control-step-delay 1 ^vt-opcode SR0
 ^allocated 0 ^maximum 999 ^group SHIFT)
(db-operator ^hw-opcode SR1 ^control-step-delay 1 ^vt-opcode SR1
 ^allocated 0 ^maximum 999 ^group SHIFT)

(db-operator ^hw-opcode SRD ^control-step-delay 1 ^vt-opcode SRD
 ^allocated 0 ^maximum 999 ^group SHIFT)
(db-operator ^hw-opcode SRR ^control-step-delay 1 ^vt-opcode SRR
 ^allocated 0 ^maximum 999 ^group SHIFT)
(db-operator ^hw-opcode SL0 ^control-step-delay 1 ^vt-opcode SL0
 ^allocated 0 ^maximum 999 ^group SHIFT)
(db-operator ^hw-opcode SL1 ^control-step-delay 1 ^vt-opcode SL1
 ^allocated 0 ^maximum 999 ^group SHIFT)
(db-operator ^hw-opcode SLD ^control-step-delay 1 ^vt-opcode SLD
 ^allocated 0 ^maximum 999 ^group SHIFT)
(db-operator ^hw-opcode SLR ^control-step-delay 1 ^vt-opcode SLR
 ^allocated 0 ^maximum 999 ^group SHIFT)
(db-operator ^hw-opcode SLI ^control-step-delay 1 ^vt-opcode SLI
 ^allocated 0 ^maximum 999 ^group SHIFT)
(db-operator ^hw-opcode SRI ^control-step-delay 1 ^vt-opcode SRI
 ^allocated 0 ^maximum 999 ^group SHIFT)
(db-operator ^hw-opcode 0 PAD ^control-step-delay 0 ^vt-opcode PAD0
 ^allocated 0 ^maximum 999 ^group WIRING)
(db-operator ^hw-opcode SPAD ^control-step-delay 0 ^vt-opcode PADS
 ^allocated 0 ^maximum 999 ^group WIRING)
(db-operator ^hw-opcode FREAD ^control-step-delay 0 ^vt-opcode BIT-R
 ^allocated 0 ^maximum 999 ^group WIRING)
(db-operator ^hw-opcode FWRITE ^control-step-delay 0 ^vt-opcode BIT-W
 ^allocated 0 ^maximum 999 ^group WIRING)
(db-operator ^hw-opcode AREAD ^control-step-delay −1 ^vt-opcode WORD-R
 ^allocated 0 ^maximum 999 ^group WIRING)
(db-operator ^hw-opcode AWRITE ^control-step-delay −1 ^vt-opcode WORD-W
 ^allocated 0 ^maximum 999 ^group WIRING)
(db-operator ^hw-opcode GFREAD ^control-step-delay 0 ^vt-opcode GBIT-R
 ^allocated 0 ^maximum 999 ^group WIRING)
(db-operator ^hw-opcode GFWRITE ^control-step-delay 0 ^vt-opcode GBIT-W
 ^allocated 0 ^maximum 999 ^group WIRING)
(db-operator ^hw-opcode GAREAD ^control-step-delay −1 ^vt-opcode GWORD-R
 ^allocated 0 ^maximum 999 ^group WIRING)
(db-operator ^hw-opcode GAWRITE ^control-step-delay −1 ^vt-opcode GWORD-W
 ^allocated 0 ^maximum 999 ^group WIRING)
(db-operator ^hw-opcode DELAY ^control-step-delay 0 ^vt-opcode DELAY
 ^allocated 0 ^maximum 999 ^group SYNC)
(db-operator ^hw-opcode WAIT ^control-step-delay 0 ^vt-opcode WAIT
 ^allocated 0 ^maximum 999 ^group SYNC)
(db-operator ^hw-opcode DWAIT ^control-step-delay 0 ^vt-opcode

TIME AIT ^allocated 0 ^maximum 999 ^group SYNC)
(db-operator ^hw-opcode STOP ^control-step-delay 0 ^vt-opcode STOP ^allocated 0 ^maximum 999 ^group SYNC)
(db-operator ^hw-opcode SELECT ^control-step-delay −1 ^vt-opcode SELECT ^allocated 0 ^maximum 999 ^group BRANCH)
(db-operator ^hw-opcode ENDSEL ^control-step-delay 0 ^vt-opcode ENDSEL ^allocated 0 ^maximum 999 ^group BRANCH)
(db-operator ^hw-opcode DIVERGE ^control-step-delay 0 ^vt-opcode DIVERGE ^allocated 0 ^maximum 999 ^group BRANCH)
(db-operator ^hw-opcode MERGE ^control-step-delay 0 ^vt-opcode MERGE ^allocated 0 ^maximum 999 ^group BRANCH)
(db-operator ^hw-opcode ENTER ^control-step-delay −1 ^vt-opcode ENTER ^allocated 0 ^maximum 999 ^group CONTROL)
(db-operator ^hw-opcode CALL ^control-step-delay −1 ^vt-opcode CALL ^allocated 0 ^maximum 999 ^group CONTROL)
(db-operator ^hw-opcode PSTART ^control-step-delay −1 ^vt-opcode PSTART ^allocated 0 ^maximum 999 ^group CONTROL)
(db-operator ^hw-opcode LEAVE ^control-step-delay −1 ^vt-opcode LEAVE ^allocated 0 ^maximum 999 ^group CONTROL)
(db-operator ^hw-opcode RESTART ^control-step-delay −1 ^vt-opcode RESTART ^allocated 0 ^maximum 999 ^group CONTROL)
(db-operator ^hw-opcode RESUME ^control-step-delay −1 ^vt-opcode RESUME ^allocated 0 ^maximum 999 ^group CONTROL)
(db-operator ^hw-opcode TERMIN ^control-step-delay −1 ^vt-opcode TERMIN ^allocated 0 ^maximum 999 ^group CONTROL)
(db-operator ^hw-opcode NOOP ^control-step-delay 0 ^vt-opcode NO.OP ^allocated 0 ^maximum 999 ^group NOOP)
(db-operator ^hw-opcode UNDEF ^control-step-delay 0 ^vt-opcode UNDEF ^allocated 0 ^maximum 999 ^group UNDEFINED)
(db-operator ^hw-opcode UNPREDICTABLE ^control-step-delay 0 ^vt-opcode UNPREDICTABLE ^allocated 0 ^maximum 999 ^group UNPREDICTABLE)
(db-operator ^hw-opcode PARITY ^control-step-delay 1 ^vt-opcode PARITY ^allocated 0 ^maximum 999 ^group PARITY)
(db-operator ^hw-opcode IS.RUNNING ^control-step-delay 0 ^vt-opcode IS.RUNNING ^allocated 0 ^maximum 999 ^group IS.RUNNING)
(db-operator ^hw-opcode INC ^control-step-delay 1 ^vt-opcode INC ^allocated 0 ^maximum 999 ^group INC)
(db-operator ^hw-opcode DEC ^control-step-delay 1 ^vt-opcode DEC ^allocated 0 ^maximum 999 ^group DEC)
(db-operator ^hw-opcode CLEAR ^control-step-delay 1 ^vt-opcode

CLEAR ^allocated 0 ^maximum 999 ^group CLEAR)
(db-operator ^hw-opcode MUX ^control-step-delay 0 ^vt-opcode MUX
   ^allocated 0 ^maximum 999 ^group MUX)
(db-operator ^hw-opcode DEMUX ^control-step-delay 0 ^vt-opcode
   DEMUX ^allocated 0 ^maximum 999 ^group DEMUX)
(db-operator ^hw-opcode BUS ^control-step-delay 0 ^vt-opcode BUS
   ^allocated 0 ^maximum 999 ^group BUS)
(db-operator ^hw-opcode ALU ^control-step-delay 1 ^vt-opcode ALU
   ^allocated 0 ^maximum 999 ^group ALU)
(db-operator ^hw-opcode CONCAT ^control-step-delay 0 ^vt-opcode @
   ^allocated 0 ^maximum 999 ^group WIRING)
(fold ^type ARITHMETIC ^prox 0.5 ^cost 0.5)
(fold ^type LOGICAL ^prox 0.5 ^cost 0.75)
(fold ^type REGISTER ^prox 0.5 ^cost 0.5)
(max-delay-per-control-step ^delay 50)

Appendix C
WORKING-MEMORY SCS

The following working-memory listing shows the decode loop from the ISPS example in Figure 18 translated into SCS as in Figures 20 and 21. Parentheses are used to delimit the boundaries of working-memory elements. The type of the working-memory element is given by the first name after the opening parenthesis. It corresponds to the names following the keyword *literalize* in Tables 15 and 16. The type is followed by its attribute-value pairs. By looking at the ISPS and VT descriptions you can see the direct translation of the ISPS carriers CPAGE and I into modules, ports, and links. This is the form DAA uses for the designed architecture.

(module ^id *.c2.constant ^type CONSTANT ^bit-left 4 ^bit-right 0
  ^value 0 ^opt-flag global)
(module ^id controller0 ^type CONTROLLER ^bit-left 0 ^bit-right 0
  ^opt-flag arch)
(module ^id r2.register ^type REGISTER ^bit-left 0 ^bit-right 4 ^opt-flag bound)
(module ^id r3.register ^type REGISTER ^bit-left 0 ^bit-right 11 ^opt-flag bound)
(module ^id v6.input.mux ^type OPERATOR ^bit-left 11 ^bit-right 0
  ^value MUX ^opt-flag global)
(module ^id v6.register ^type REGISTER ^bit-left 0 ^bit-right 11 ^opt-flag global)
(module ^id v6.x4 ^type OPERATOR ^bit-left 11 ^bit-right 0 ^value @
  ^opt-flag bound)
(module ^id v6.x7 ^type OPERATOR ^bit-left 11 ^bit-right 0 ^value @
  ^opt-flag bound)
(mux-port-count ^module v6.input.mux ^icnt 2)
(port ^id *.c2.output ^type OUTPUT ^bit-left 4 ^bit-right 0 ^module
  *.c2.constant)
(port ^id controller0.source0 ^type INPUT ^bit-left 0 ^bit-right 0
  ^module controller0)
(port ^id r2.output ^type OUTPUT ^bit-left 4 ^bit-right 0 ^module
  r2.register)
(port ^id r3.output ^type OUTPUT ^bit-left 11 ^bit-right 0 ^module
  r3.register)
(port ^id v6.input ^type INPUT ^bit-left 11 ^bit-right 0 ^module
  v6.register)
(port ^id v6.input.output ^type OUTPUT ^bit-left 11 ^bit-right 0
  ^module v6.input.mux)
(port ^id v6.input.source0 ^type INPUT ^bit-left 11 ^bit-right 0 ^module
  v6.input.mux)
(port ^id v6.input.source1 ^type INPUT ^bit-left 11 ^bit-right 0 ^module
  v6.input.mux)
(port ^id v6.x4.output ^type OUTPUT ^bit-left 11 ^bit-right 0 ^module
  v6.x4)
(port ^id v6.x4.source0 ^number 0 ^type INPUT ^bit-left 4 ^bit-right 0
  ^module v6.x4)
(port ^id v6.x4.source1 ^number 1 ^type INPUT ^bit-left 6 ^bit-right 0
  ^module v6.x4)
(port ^id v6.x7.output ^type OUTPUT ^bit-left 11 ^bit-right 0 ^module

v6.x7)
(port ^id v6.x7.source0 ^number 0 ^type INPUT ^bit-left 4 ^bit-right 0 ^module v6.x7)
(port ^id v6.x7.source1 ^number 1 ^type INPUT ^bit-left 6 ^bit-right 0 ^module v6.x7)
(controller-port-count ^module controller0 ^icnt 1)
(link ^source-port *.c2.output ^source-bit-left 4 ^source-bit-right 0 ^dest-port v6.x4.source0 ^dest-bit-left 4 ^dest-bit-right 0)
(link ^source-port r2.output ^source-bit-left 4 ^source-bit-right 0 ^dest-port v6.x7.source0 ^dest-bit-left 4 ^dest-bit-right 0)
(link ^source-port r3.output ^source-bit-left 6 ^source-bit-right 0 ^dest-port v6.x4.source1 ^dest-bit-left 6 ^dest-bit-right 0)
(link ^source-port r3.output ^source-bit-left 6 ^source-bit-right 0 ^dest-port v6.x7.source1 ^dest-bit-left 6 ^dest-bit-right 0)
(link ^source-port r3.output ^source-bit-left 7 ^source-bit-right 7 ^dest-port controller0.source0 ^dest-bit-left 0 ^dest-bit-right 0)
(link ^source-port v6.input.output ^source-bit-left 11 ^source-bit-right 0 ^dest-port v6.input ^dest-bit-left 11 ^dest-bit-right 0)
(link ^source-port v6.x4.output ^source-bit-left 11 ^source-bit-right 0 ^dest-port v6.input.source1 ^dest-bit-left 11 ^dest-bit-right 0)
(link ^source-port v6.x7.output ^source-bit-left 11 ^source-bit-right 0 ^dest-port v6.input.source0 ^dest-bit-left 11 ^dest-bit-right 0)

Appendix D

THE SYSTEM/370 CRITIQUE

This appendix lists relevant sections of the IBM System/370 critique by Claud Davis. Also present were Ralph Bahnson, Ted Kowalski, Sumit Dasgupta, and an unknown IBM employee. The font convention followed in this appendix displays Davis's remarks in *italic*, and Bahnson's, Kowalski's, Dasgupta's and the unknown IBM employee's remarks in roman, with the speaker's name in **bold**. If the speaker's name is omitted, the speaker is either Davis or Kowalski, depending on whether the font is *italic* or roman.

D.1 Getting Started

Claud: *"Well, the few I constantly worked with were Roy Simmons and Ray Warren and, let's see, of course, Jerry and I had another guy, but there were only five of us. Tom Wart spent a lot of time with us. That was about my whole team."*

Ted: "There was a team of five or four?"

"How's that?"

"There was a team of about four who did this?"

Claud: *"Yes, but we did have support of EDS, tools — that sort of stuff."*

Ted: "The system that I have — what I'm going to talk about today in the talk is that system — The level it takes it to is not really down to the — your mentioning 700 transistors or something or circuits only use 5000 — it takes it not down to that level, but what it's trying to do is create an architecture and so it looks at a whole description for an IBM 370 and says, 'OK, let's see what we think there should be in terms of registers, what we think there should be in terms of ALU's, what kind of bus structure, you know, how do we get information from one place to another and for each one of the data paths it generates a microsequence tablet or a controller for a microrom — or something — that each one of these lines can be turned on in these clock phases.'"

"Yes."

"This is all up to the technology independent level (like you say), and it doesn't know anything about how you can actually implement a register. It just says, 'Now you can get a register that "a number of bits" or that incrementing or decrementing, shifting' — something like that."

"OK, I understand that. We also went through that phase as well, but we were constrained by the physical things that we had to work with, saying, 'Yes, we're going to have registers but when we add things up, it just fell off the edge of the chip. Now, what are you going to do next? And so, that is when we come back and did what I call brute force: chopping here and there. We took out registers that we would like to have kept, and I'm not expecting that kind of sophistication necessarily, but we eventually will put in that optimization, I'm sure."

"Sure."

"Or you can optimize for speed, you'll optimize for power, or you'll optimize for cost. Those are normally the three things that you go after or at least we do."

"My system does optimize for those sort of things in a crude sense. I don't have the bottom level numbers to bubble back up but what I say is, 'Well, if you want to optimize for space, then you have to control about how many ALU's you can have. You can have control about how many adders, how many registers and you can limit those things. If you want to optimize for speed, you can say this critical path has to be done in this many clock cycles, given that you know how roughly if you're doing it in a standard cell sort of fashion you say, well, OK, I know an AND of two things takes a number of microseconds and my clock is 50 mics, so then I'd know how to stack those things. I can say I roughly at the top level sort of know about that stuff.' It's only a first level approximation, but you got to see what would be nice to do is, of course, take the system to the next level down, the next level down and bubble that system altogether."

Claud: *"OK, we went through two phases. It sounds like you are capable of handling the first one very well, and then when we start trying to take that, map it against the physical world, why then we make additional tradeoff. So you took the description of 370 and from there went for register layoff — what I'll call a register transfer design."*

Ted: "Right, exactly."

D.2 The Talk

Sumit: "OK, I guess we should get started."

Ted: "Yes."

Sumit: "Jerry will get caught up to us one of these days?"

"Yes, yes, he's lost somewhere as usual. Kill the lights, I guess. That'll make it a little easier."

Ted: "Help me move this back a little."

"OK, today I'm going to talk about the research that I've been doing for my thesis. I'm doing this research with Don Thomas. He's my advisor. This research is being done at Carnegie-Mellon University. The title of the talk is 'A VLSI Design Automation Assistant.' It's titled 'Assistant' for a good reason in that it is not something that goes through and does everything automatically but rather it's meant to sit there and aid the designer — indeed, some of the comments we were having before about: 'Gee, once I get to this level, I can go down to the next level down and then I can start making decisions about what I want to pitch, these registers, I don't want to keep them as architectural registers.' Those decisions can then be put back up in the system and run again, and then the system will come out with a different design."

"This morning I'll talk about several things. I'll state the problem I'm trying to attack and my research goals (what I'm trying to get out of it). I'll speak a bit about other research that's been done in the area, since as you know no research really stands alone and how the other research has motivated my research. I'll give a brief introduction to knowledge based expert systems so we can all talk the same language. I'll talk a bit about

Section D.2   The Talk   129

how acquisition interviews have gone in the past and indeed we're going to do one today. I'll show the sort of things I've typically learned from these sort of interviews, and I'll talk about where this research is going."

Ted: "Basically my problem is to start out with trying to do synthesis. As you all know in the next five years to ten years, we're looking to make our designers much more efficient, and we're not only looking to gain productivity from designers, but we're also looking to do more and more designs. And since the number of designers doesn't look like its increasing by any great numbers, we want to be able to take the designer's knowledge and spread this out to a number of people, so a number of people who don't have the knowledge of an expert designer, such as Claude here, can actually go and do a design. So we try to do synthesis. Now, what does that mean? Well, we want to describe the problem - describe the thing we want to design in the algorithmic sort of sense or behavioral sense and that's what this description is here. It's a standard programming language, the language I happen to have chosen is ISPS (that's because this is the language we have at Carnegie-Mellon and it is used by a lot of people in the VHSIC project and such. It just has conditional statements in which you can do if/thens, and you can do decodes; so if the value of pb is 0, or 1 do a couple of things, and it has concatenate operators, it has bit width specifications, standard sort of ALGOL-like language stuff with the additional ability of specifying directly on bits and concatenations. From that language, what I produce, in a pictorial sense is this sort of thing. This actually is the 8 bit data paths of the 6502 without allowing busses. This is just allowing a multiplexor style design. So, what we go from is something that we can test and simulate to something that produces canonical registers in these boxes, canonical single operators such as AND, and SHIFTs and NOTs, ALU's (that's another register and central accumulator and memory, and the data paths between them. Also, what we have is a symbolic microsequence table or a symbolic microcode that we produce so that we can optimize that, compile it and turn it into a microengine that will control each one of these lines so we'll know which lines to turn on, turn off, and how to strobe the registers and how to select functions. The little things in the registers here are the bit width and like some registers, can do shifting left and right, some registers can do increment/and decrement; all these things are found by the system automatically from the description."

Claud: *"What complexity clock can you handle? You use the term 'clock' I gather for gating registers."*

Ted: "Right."

Claud: *"OK, four phase or something — symmetric or asymmetric?"*

"The rules I have are for symmetric clocks."

"Symmetric?"

"Typically two phase."

"OK."

Sumit: "I notice, excuse me. In some of those blocks you have two outputs like say, this block right here, is it true and compliment?"

Ted: "No, this is just a fanout from that. This actually is an accumulator. The inputs are all on this side; outputs are on that side, and the accumulator is actually going into the input of that ALU and also into the memory blocks. Though I've learned a lot about how the MOS designer seems to do designs, I've learned not very much about drafting at all. It's sort of a pity. You can see from the 370 we don't have tools to draw the stuff out yet except by hand, so we do all the stuff by hand."

Sumit: "Yes."

Ted: "And that's a pain in the neck. I've got of a program which would assist me in doing that for small designs, but it only handles single page things. I'm afraid it doesn't handle very big things yet, and we have someone who's actually working on doing the whole DP package at CMU on our HP9000 series computer, and that has a bit display and we'll be able to do the stuff. In fact there's an interesting error in this design."

"Yes, I see a connection is not made up there on ... What is it A — or? No, the one there to your left. Right there."

"No, that's a constant one that's coming into the shifter."

"Oh."

"I'm sorry. But the interesting error is here. If you'll look at this AND, and that AND, we'll see that this AND gets an input from the same place that that input ... [comes from] and that AND gets an input from the same place that that does. So those two ANDs shouldn't really exist. There should only be one AND in reality. And the error was that when I had drawn this and shown this to designers before (I was going through the acquisition interviews) I had misdrawn this. I had drawn it connected to something else, so I never saw that my system produced this incorrectly.

Indeed, what the correction was I had a rule that said that if an input to source 0 (which is something) and you input the source 1 (which is something) and it goes out to someplace, and it has something else which is exactly like that, well then you shouldn't have two of them you should only have one. That's true for subtracts and for compares but and's and or's, it doesn't matter which order the sources are in. So when I draw it out on here, I actually drew the sources in the same order, but in reality the way my system had them, it had them reversed. So I never noticed that. It's not a whole together thing, but it's ongoing research as most research is. So my goals and my research are to produce a system that can go from an algorithmic system to an architecture. To do this I have to do two things: 1) I have to extract knowledge from expert designers. In particular, I have chosen designers who have been microprocessor designers and MOS designers. My system has a great deal of knowledge about how to do MOS microprocessors, but it doesn't do other things very well. If you were trying to design a disk controller or floating point unit, it has default rules to use. It would, indeed, it would produce a design which worked, but wouldn't be as optimal or as optimized because there's no special knowledge in the system for that."

Unknown: "Why do you make the distinction between MOS and BIPOLAR at this very high technology independent level?"

Ted: "In a sense I really don't in my system. In another sense since all of the designs I have used have been in this category. I know the architectures that they choose ..."

Unknown: "The redundancy removal operation that you finished explaining on the previous page? Could be there for technology reasons?"

Ted: "Sure."

Unknown: "Probably not at this level."

Ted: "Right."

Unknown: "Not assuming distinctions between MOS and BIPOLAR."

Ted: "Right. Now it is just to say who have I used for my experts, and that's indeed whom I have used."

Unknown: "OK."

Claud: *"At the level you are, it really doesn't matter anyway."*

Ted: "No. It really doesn't matter."

Claud: *"It's at the technology level that they make ..."*

"What does matter is microprocessors. All the designers are really aiming for and really trying to do are microprocessors, are controllers and ..."

"There is one distinction in the event you are working with MOS that would be different and that would be in the implementation of pass gates and that sort of thing. We don't have that in BIPOLAR."

"Sure."

"But at the level that you are right now?"

"That doesn't matter."

"That would just show up as a connection."

"Sure."

"Of some sort."

"Sure, that doesn't matter at all. And the second thing that I do, once I claim to have extracted knowledge from an expert, is implement a system to goes and test it, so you can say that I did a bunch of interviews, and that's nice but I can't prove that the knowledge works unless I have a system. So I've gone and written a system and indeed in writing the system and adding the knowledge in it, it's a way of debugging the knowledge and we do iterations with experts so we add more knowledge. Often in an iteration cycle, an expert will tell you something. He or she says, 'Well, this is true.' And then he or she will come back two days later and say, 'Well, that's not really what I meant. What I really meant is that's true only in this case.' And so there's a context that you don't always get from the expert. You see, experts are really experts at doing VLSI designs. Or they're experts at doing whatever they're experts at. But they're only novices, at explaining what they do. They just say, 'Oh, what I've done. I've done that for years. I don't know how I do it. I see it; I recognize it.' So the paradigm I'm trying to capture is that humans are recognition engines. They'll look at a problem, they'll recognize a problem and because they see something in the problem they know how to solve or they solved before they do it with the same solution that they did before. Often they start from past designs and they start muddling from there or they go along and they'll find something: 'Ah! I remember I did this, yes, two years ago. I did this on this other processor.' And that whole piece will come back. So those sorts of recognitions are what I try

Section D.2 *The Talk* 133

to code in rules."

Ted: "Mr.?"

"Sure."

Unknown: "Do you allow the user to change or update the rules at all?"

Ted: "No, not at this present time. There are two very difficult issues here: one is the domain that being VLSI and the second is a hot topic in AI itself these days and that's the knowledge acquisition — how do you acquire knowledge, how do you add knowledge to the system, can you get the expert closer to the system, can you get rid of the knowledge engineer and, indeed, the research that I'm hoping to do when I'm done with my thesis is carry on with the knowledge acquisition problem because I don't think that someone like myself should spend a year of their time, two years of their time working with experts. When I've dealt with experts (the VLSI designers) a number of them after they tape out on a the design say, 'OK, well, what I'm going to do is make a tool to go and do it so I don't have to go through that trouble again.' Well, I've gone through the trouble of doing knowledge acquisition. I have some ideas I'd like to work on when I'm done with this. I'd like to step back and try to implement a system to help experts actually directly through acquisition. But that's not done yet."

Claud: *"What I look for, very frankly, in tools is something that will handle the detail. I have difficulty in realizing, all right, that I've used 86 pins; I've only got so many left. I'd like to be able to know that as I move along without having to stop and count and discover I've overflowed and disturbed things. And you just make very trivial mistakes — the ones that kill us — it's not the fact that ..."*

"The big ones, too. You have the inverse of what you wanted or something simple like that."

"For example, on our first implementation we had four mistakes: one of them was an inverted line, the other was a redundant line, which you can't test and it's these kinds of things."

"Sure. It's a game of — the metaphor I like to use is, it's a game of juggling, as a person can juggle so many balls. Some people can juggle four balls; some people can juggle five balls ... to use a tool you're adding another ball to your juggling. The tool has to at least remove one of the balls that you're juggling so you can maintain status quo, hopefully remove the second ball so you can juggle less. Then you can add something else.

That's the game I think I've provided with this system."

Ted: "Other research that's going on in the area ..."

Claud: *"Before you proceed, you say you go back and implement, you don't physically implement or do you simulate?"*

"We use a software simulator."

"You use a software simulator rather than hardware?"

"Right."

"OK."

"The other research that's going on that started this research has gone on at Carnegie-Mellon and also in Germany. First, I guess it was about four years ago or five years ago, we had a CMU register transfer level synthesis that was done by a fellow by the name of Lou Hafer, and that was a pretty classical brute force method of doing synthesis. It had a description starting with ISPS, and for every ISPS operator for every line of ISPS, there was hardware that was implemented. So if you had in your page of descriptions five places where you get addition, then it would by God have five adders come out. If you declared 12 registers, then it would have 12 registers. I mean it just strictly did a transformation."

"That would be an assembler rather than a compiler."

"Yes. Actually a number of the silicon compiler things that are going on these days, quote unquote, are simply that same sort of thing. If you say it, it's going to do it, period."

"Line for line I call it an assembler."

"Yes, not very elegant. More elegant work was done in Germany a project called Mimola, also an assembler, but this assembler was more interactive; you could sit there and poke at and you could limit the number of registers saying I only had 'n' number of registers, can you help me with that. But it was very time consuming, and the designer has to sit down and implement the behavioral description in the optimum parallel fashion first. Humans are pretty bad at guessing parallelism, so there are a lot of drawbacks to that stuff. Another piece of research was done at Carnegie-Mellon (also by Lou Hafer for his doctoral thesis), and that was doing the constraint language approach to the problem. He said, 'OK, the way we'll do it, we'll use a mixed integer linear programming problem approach, and we'll put all kinds of constraints, all kinds of equations on the design, and

Section D.2   The Talk   135

we'll solve them all simultaneously and they'll come out with the optimal design for the constraints you've given.' Unfortunately, the problem blew up. As you know, with humans there are a lot of constraints with the VLSI design. For very simple circuits like circuits of five or six gates and a couple of transfers, there were, I think, many equations that had to be solved simultaneously, and it just took forever. It couldn't handle anything bigger than a few gates."

Claud: *"Oh, I can believe that."*

Ted: "Sure, we do this all the time. We sit there and say, ...' But we know, you see, one of the things as a designer, we know what constraints to sweep under the carpet. We know not to look at that constraint right now. We don't care about that. We'll look at that at a better time, so ..."

"In dealing with a master slice or a gate array design you're actually solving the equations correctly."

"When you're down in silicon. Oh, it is worse ..."

"You need disk packs plural for a single chip ..."

"A single chip, sure."

"Well, you need a good engine there."

"So, that isn't the way we want to go. Another system was done for ..."

"30 hours just to give you a feel for it and we were playing with gate array."

"No, it was system time 30 hours. We actually didn't run that long."

Ralph: "But that wasn't for the actual devices ..."

"Well, we'll also take the actual devices and we'll take those and run those over the weekend. So, I believe you've got on a, you know, on a five or ten MIPS machine you got days to run it. So that's out."

Ted: "So, that wasn't a good way to go, so the next thing that was tried was doing local minimax sort of functions for the algorithm where can I within my neighborhood locally put this so it won't hurt me very badly, and this is still a NP problem; it still has exponentials but not as bad as this, and it produced designs that are locally optimized but in a sort of global sense it didn't really map what humans do. It seems that humans sit down and say, 'Well, this is the general plan, here is the general layout of the chip, or here's the general notion I think I need.' It had no notion of this general road map scheme, but it does produce working designs. OK, so as I said before what I'm trying to do is use knowledge based expert systems to see if that technique can be used for the VLSI design. Well, what is the technique? Let's talk about it a bit. It was the technique I mentioned earlier. It's the technique of seeing something in the current problem that you remember from before and then doing something. See, you have a bunch of 'if this and this and this and this and this is true,' then do something. In particular, I'm using a language called OPS5 which is the language developed at Carnegie-Mellon, but that's not really that important. What's really important is it gives you the features of just putting in rules and having the system actually choose when to implement the rules. So you divide the problem into several pieces: you divide it into a blackboard and on that blackboard, which is called working memory, you put the problems. And all the rules can look at that blackboard. And you have a rule memory as a second memory. That memory has a bunch of these 'if/then' sort of conditions. And you have an inference engine, something that runs, looks at the rules, looks at a subset of the rules, determines which rules are applicable to a problem, and chooses one of those rules. Now, you might ask well, but gee, there's hundreds of things I could do at this point in time. How do you know which rule to use? Well, it uses several disambiguating techniques. One, if you have two rules and one rule matches something more current in working memory, use that rule. And the reason why it does that is because humans who are working on a problem will continue to work on the same problem until they can't go any further on that problem. The other rule they use is if two rules match the most current thing, but one rule has more knowledge than the other rule (in other words one rule is a special case of another) then use a rule that matches this special case. The reason for this is because why, as a human you can say, 'Well, gee, in a general sense, I do this but, ah!, in this case because this other thing is true I would do that instead.' And so, it tries to map how humans would do this sort of thing. It provides the ability of doing that."

Ted: "Now the advantage to doing this sort of system is that if I was coding this in a programming language in PL1 or C or whatever your programming language is, I'd have to go and know where in my system to put each one of these 'if' statements. I'd have to know where each 'if' statement had to be evaluated. Well I don't have to do that because the rule interpreter dynamically picks it based on these two disambiguating things I just mentioned. Well, what does working memory look like? From my system here are just a couple of the thousands of elements sitting

Section D.2 *The Talk* 137

in working memory, I have basically modules and ports and links, and this is how I describe my technology-independent world, where module can be one of several things. It can be a register, a memory, and operator, constant, or controller. A module has fastened to it ports. A port is where you can make the connection either in or out or by directional, and then, of course, from a port you have a link that links to something. These two elements describe an 18 bit adder for two's complement addition, and this is the input port which is also 18 bits; it's associated with the adder and there's nothing very magical about it. So, this is just a way of keeping track of things internally, but the interesting thing about expert systems is how rules are written. This is an English version of the rule, and I can show you what the rule really looks like. It looks more like the language called LISP which has parenthesis and such, but this rule is something we learned a very long time ago in basic design."

Ted: "The notion is: you have a connection to something, and you want to make another connection. Well, we all know that you can't make two connections to the same place or else you've now OR'ed those two lines together. But in reality what you have to do is put a multiplexor in the middle, so you bring one line into the multiplexor, another line into the multiplexor, take the multiplexor and connect it. Here we recognize that what we want to do is make a link and there is another link to the same place already, and this link is on a multiplex already, then what we have to do is make a multiplexor, and we have to hook the two things into the multiplexor and connect the multiplexor to where you want it to go."

"My system is composed of 283 rules which are very, very simple. This is not a difficult concept. We have a bunch of little rules and they all sort of work in clusters, so if you want to change a rule about how multiplexors are done, then you just change one of a couple of rules. It's very easy to add new knowledge to the system. If you want to just add knowledge about new recognition like — well, gee, this optimization about switching the two changes on the operators like AND's and OR's and such, I just dump another rule in; I didn't have to look to the rest of the system."

Ralph: "By the way, I don't know if you realize this or not but you just did tie yourself to MOS. When your wires come together, they create an OR. Wired OR's are a MOS related thing. If you wire together in BIPOLAR that connection seems like an AND."

Claud: *"It can be either."*

Ralph: "In current switch, the point is that it is a technology related rule."

Ted: "This rule avoids technology-related issues, because if I put a multiplexor in the middle, then I don't have a problem in either case."

Ralph: "There [is] context that is not stated is what the point is."

Ted: "And then that is indeed why I come up front and I say look, I dealt only with MOS designers. If things come out that look strange it's because [it's not BIPOLAR] (as I often tell the story), gee, before I started this game, I could go to a cocktail party and I could be a computer scientist and people would know that I am a computer scientist because I'd sit and talk shop about programming languages and such. Now I go to a cocktail party and be a VLSI designer and for that sort of level, a party can get by quite well. And I could talk shop about VLSI design and they would never know. So I've learned. That's the sort of knowledge I've learned from. People you hang around with, I guess."

"What does my system do? This system has several temporal-ordered tasks that it does. These tasks came about from a series of interviews I did preliminarily with designers saying, 'OK, sit down with a designer and say what do you do? How do you design.' We sit and talk about design, and from this I got a notion that they did several things, how they looked at the world, saying, we want to get a global picture. We first got a hold of the global inputs and outputs and how to deal with that. What kind of constraints do they have? What do you want me to do? And then they carried out to partition the problem into pieces and how they did partitioning. Eventually, once they had a piece and how they actually did that piece and within the piece they said, 'Well, the first thing we looked at was assigning things like memories and ports to the outside world — things that you said were architectural registers, and you don't want to have go away.' I call that my base variable storage, so I have rules to go over and determine what things are declared as architectural registers, where memories are, and assign modules to them, assign memories, output, input, address ports and things like that. Then I have a series of rules that look up and down the thing you want to implement and decide how it should be broken up into clock phases. Now this is sort of a temporary thing. As you know when you assign clock phases that sort of decides what registers you have, but then as you bump into a constraint, you can reorder your clock phases a bit and say, 'Well, gee, I can take this thing that I have (this thing that I've done very parallel) and I can serialize it.' So I have rules that know how to serialize and push back and forth; it goes in either direction. Once I've assigned clock phases, I take the description that I

Section D.2 The Talk

have, this ISPS description, this behavioral description, and I map it into my world of registers and modules, ports and links. I call them temporary because I don't really assign them yet. All the rules that I've gathered from experts who have looked at papers and said, 'Well, gee, this module in this context should go away.' Because all my rules are written like that, I map everything to that world first and then I have my rules which actually do the system. So that's the temporary phase. Then these things are done both at the same time. I'm assigning registers, I'm assigning modules, I'm rearranging clock phases, I'm keeping track of my constraints, and that's how my system runs. Then it goes back and does the next partition of designs. It does it over and over again."

Claud: *"Just so I'm sure I understand you, a module? Maybe I was thinking ... When you explain that. What do you define as a module?"*

Ted: "A module can be one of a whole bunch of different things. It can either be a register, it can be a memory."

"A functional unit?"

"It's a functional unit."

"OK." You see, we have a very specific meaning for it."

"Oh, I'm sorry."

"You understand?"

"Yes."

"And even though I know better, I wanted to be sure I knew what you were saying."

"Sure."

Sumit: "Will it be possible to get copies of the presentation, please?"

Ted: "The talk I gave at the SRC talk which Ralph has a copy of already, and I will be happy to send you the slides I have from the DA conference which are an update of this. The system now, too, has added rules for busses, and I can send you those if you like those, too."

Sumit: "Well, as far as the DA conference, we'll have plenty of proceedings laying around."

Ralph: "Right."

Claud: *"In fact, we'd be better off having stuff that's in the public domain anyway."*

Ted: "Right. You have the SRC's so far?"

Ralph: "We have the copies."

"These are all public domain?"

Ted: "Yes."

"OK, how do I deal with getting knowledge into the system, which addresses the question you brought up earlier? It's difficult. It's not an easy thing, and I tried a number of different approaches for extracting knowledge from experts. I tried giving them a verbal description of things. Here's a printout that says what modules are connected to what — can you help me?. That was a bad idea. I tried giving them an interactive system in which they can type commands and they would draw things from them. And they got so involved with typing commands and trying to remember the command syntax, that was a loser. But the thing I finally ended with and it was pretty good was to go and take a drawing, put it out on a table and then take the drawing and cover it up with a piece of plastic. And at first I was also using plastic and cardboard, so when I first looked at the design and then I would cover it up with cardboard, and every time they wanted to look at something, they'd pick up a piece of cardboard and then they'd look at it and tell me something about it. And if they'd tell me something about it, they'd take a marker and write on the plastic. If they wrote on the plastic, I'd put down another piece of plastic, so that I'd build up these piles of plastic and I'd keep them covered up with cardboard. So I got a protocol of what they were looking at, what they were focusing on, and what changes they made. And I also had a tape recorder running, and I'd tape the whole thing and transcribe the whole thing later. Actually, I never really transcribed these tapes; I merely played them back and listened to them. I took notes later so that I could be involved with the design. That's a good technique. It turned out that the cardboard is probably a little bit overkill, because I often find a few who'll go, 'Lift, lift, lift, lift. Oh, that's what I want.' They play the game like — what is it — Jeopardy on TV where you couldn't remember what cardboard it's behind. So I've since stopped using cardboard. We just look at the design, and I've gotten better at learning how to ask the right questions. Before I didn't know how to ask the right questions well. So that's the technique that I use, and that's the technique we use today."

Section D.2   The Talk   141

Ralph: "Basically, are we going to have structural transformations?"

Ted: "Structural transformations. Right. I wouldn't do this sort of thing in an architecture. I'd rather do this."

Ralph: "How about looking into getting them to write pseudocode at a very high level if I have a box that is connected ..."

Ted: "That's what I do verbally with them."

Ralph: "To the same box with identical function then I can remove one of those boxes and connect them."

Ted: "That pseudocode is what I actually have in the system (in the expert system). I don't have them write. If you look at the psychology literature, you can break people up in at least 2 ways. One, is people who are better at verbal skills and one who has better visual skills. It turns out that the VLSI designers seem to be better visually oriented. It seems to be better to say, 'Circle this and do this.'"

Ralph: "That's true but you miss out on the little details. You have to put it down in a formal algorithm. Then you remember to ask: What happens in this case?"

Ted: "Well, that's indeed what ends up happening. I codify that. OK, dump it into the system, and then they see the results of it. What you really want to do is, in the best of possible worlds if you have very fast computers, you can say, 'Oh, gee, I don't like this design and if you see this do this,' then you want to see your whole design change in front of you by that rule and see what kind of effect it had everywhere. Because sometimes it will have an effect where you didn't mean for it to have an effect. But the problem is that this system to do the 6502 design is three hours CPU time. So you can't expect an expert to wait that long. To do the 370 design was like 50 hours of CPU time. To get the time of [designers like] Claude here is difficult. Their time is so valuable, they just want to give me this, this, this and run off. That's fine because I'm grateful for every little bit that I can get so that's the best way I've found to optimize my time. If I had an expert who could spend several months on the problem, then I would indeed attack it a bit differently. But I think that's a real good suggestion. We're going to have ..."

"There's a language called OPS83 which will improve my performance quite a bit on this system. OPS5 is a language which is written in lisp. Lisp is a compiler and interpreter on our VAX's. It itself is interpreting the rules. Going from this system to OPS83: OPS83 takes the rules and produces VAX assembly code from them directly. I'm hoping to get about 30 to 1 to 40 to 1 speedup. So I'm cutting my time down. I'll take this IBM design (which is more or less 60 hours). I'll reduce that to two hours, so I'll be able to get a faster turnaround and then I'll look to a much more interactive system."

Ralph: "What about [the] PROLOG machine in Japan?"

Ted: "Two projects at Carnegie-Mellon are going on, implementing hardware to assist OPS-like languages, just like PROLOG language. They're just thinking about the architecture. It's too early to say anything. They don't really know. But I think the hardware system approach is really the better way to go for these NP problems. So, the sorts of designs I can produce, the things I can recognize, this is a 6502 design, canonical registers. Gee, I can recognize that you're just doing incrementing, decrementing into the registers. I don't need to have an ALU-type function to do that. If you have a register already, it easy to add the addition stuff. You don't have to have the other latches. Shifting within a register is easy. It's just wiring. I recognize how to put things together. This design had particular problems. If you look in here, there's just a bunch of junk. What are all these lines here? Well, this was the start of the set of interviews that I used for developing busses. You'll see that in the 370 design that I have a bus that tends to an ALU and register along side of it, so I carry the bus over and it takes the two-phase clock into the ALU."

Claud: *"The operations like shift. Those are linked to your original source specifications?"*

"Right."

"You don't synthesize the fact that you need to put together a bunch of ANDs."

"No. I have this behavioral description and from that I synthesize out what I need, but I don't make up the fact that, 'Oh, gee, I know how a BALR instruction works; therefore, the system knows BALR or because it's been told the BALR from the ISPS code.'"

"Other things I want to do with the system: I've designed an IBM 370. Today we're going to sit down and critique that, and other things that could be done with the system is it handles microprocessor designs. One could add rules to know about floating point designs. It's nice to go and add those sorts of things. Another thing a lot of people are doing — I was working with a character at INTEL, and he was starting work on a DSP

Section D.2   The Talk   143

project and they have processors which are very pipeline oriented. So you take an instruction and fan it out over three clocks and I have no rules that know how to deal with pipeline. I guess that would be nice to have, too. But those sort of things are for future development. There are other people who can carry on that kind of research."

Ted: "In particular, what I want to address today is the issue of the critique. The kind of things I'd like to get from the critique are two fold: one, is a general sort of sense of what do you think about the design, a sort of a qualitative kind of sense. Another sense is a quantitative sort of thing. While I want to talk about my design, I want to see how you compare designs. Do you compare it by the number of bits and registers? And that's a very mushy kind of area, and any kind of comparison numbers that I can get to later go and say, 'Look, the thing that you publish now the ICCC had registers, but there are a lot of things hidden in that little, tiny drawing, I'm sure.'"

Claud: *"Oh, yes."*

"I'd like to see if I can get more information about what was actually in that design or at least we'll talk about possibly sending out some papers or something. The sort of critique that I've done is just saying, 'OK, well, how many bits of each operator, how many bits of each register, how many bits of microcontroller do you have?' Those are pretty rough metrics, and I would be happy just to even get those kinds of metric from the ICCC thing, if we can look at that."

"In summary, I've talked about the problem itself, my research goals that of going from algorithmic descriptions to an architectural description. I've spoken of other research and how they're either unacceptable or too expensive and why that led me in this direction, I've addressed the issue of knowledge based expert systems and, in particular, talked about how it happened to divide into three pieces: a working memory (which has a problem in it), the rule memory (which has the knowledge of what to do), and the rule interpreter (which actually looks at which rules to fire). Those are pretty easy to change — specific little rules. And as a passing point, people often ask me, 'Well, gee, it seems that this stuff is very muddled and mixed together. How do you ever deal with it? When you change one rule, doesn't it have to change a whole bunch of things?' The answer is no. I've kept pretty careful notes on whenever I've made changes. The most massive change that I made was and when I added the rules to do busses, I added 35 rules to do busses. Those were in a file by itself. I only had to change two other rules. The two rules I had to change were how to deal with multiplexors. At this point in time, why didn't you do that and sometimes it could have — just chose to do the other. I spoke a little bit about acquisition interviews and how they really come up with common sense knowledge. That multiplexor rule really is just a ... You'd say, 'Well, gee, that does make sense.' And the rules are nothing magical ... They're just a bunch of common sense pieces of knowledge. And today will be the test of the 370."

Ted: "Questions? Comments to start with?"

Ralph: "Well, I have a question but does anyone else want to go first?" To sum up [you use] your ISPS description to create a naive architecture from that and then use the knowledge base system to evaluate structure transformations against that model?"

Ted: "Well, what's really done in the expert system is the following: It starts out with ISPS. I compile ISPS into an intermediate form. The form that I compile it to is called value trace and it's a data flow language. A data flow language meaning (I'll use the blackboard) ... Essentially what I compile it into is a form that makes it easy for me to go on and look at transformations and such. Let's say we have $a = a + b$ and $c = a + b$. What I want to do is get rid of what designers do to things. The person who writes the ISPS can be a person who's very cautious and uses lots of variables or could be a person who puts everything in one expression. And I want to get rid of those sorts of things. Somebody may not really need all these expressions and may find a better way of optimizing the variables. What I do is transform this into value 1, value 2, and this into value 3. And another transformation, value 4, value 1 produces value 5, where value 1 at that point in time happens to be value 3. So I take this form and I look at this. That's in the value trace. I don't really create an architecture and then change it, rather I look at this and say from this I know just what to do in the architecture."

Ralph: "But you're dealing with a structural representation of the model?"

Ted: "Right."

Ralph: "Now that you have this internal structural model, can you give a little light as to how the expert system decides to choose a particular rule. For instance, does it start at one section of the network and just tries to match the particular characteristics of the network at that point?"

Ted: "Right, it's pattern-matching."

Ralph: "It's a structural pattern match?"

Ted: "Right. I guess the word structural sort of bothers me in a sense, but yes, it is. Structural match just means a different word to me, but yes."

Ralph: "But it's dependent on a particular time you come on any node in that structure."

Ted: "Right."

Ralph: "It's kind of local in a sense. Its not taking a look at everything at once."

Ted: "No, if you had a big network of things and they're connected together in some weird fashion, it looks at the whole network all the time and it says, 'Of this whole network, all these possible rules are possibly active.' So you evaluate all the possible rules and then (this is called the conflict set), once it sets up its conflict set, then out of that it chooses one of the rules to use based on either the time itself or based on the most specific rule. No, it's not doing local things. It's always doing global things. It's always looking at the whole world."

Ralph: "You're not deciding which rule to use based on some metric of optimization?"

Ted: "No, the rules are chosen based on these two premises discussed earlier or based on a default premise as to which rule is first in the file if these two things failed. It matches patterns in the whole network. [I use] The language, OPS5, to match patterns of things (I have been matching things like: here's a module with a port and a link going to a port, going to a module), well, those are all separate working memory elements. So, I had them all linked together through variable names. I have a huge number of variables, and so to create all the possible rules to this network ... One particular match that may invoke three or four intermediate stages of matching in OPS5, I sometimes have 1000, 2000 active and current things to look at, so my expert system works slowly because of that. That's the reason why it takes three hours to do the design. Doing pattern matching, you understand, is a hard problem, and because of that all this matching takes a long time. Once I set up a match, then I just choose one, pull it out, and it looks at what local part of the network has changed as to what rules now have to be eliminated or what will have to be added and then it evaluates all the rules."

Ralph: "Could you explain once more about the time of working memory?"

Ted: "OK, each one of these working memory elements has gone into working memory at time 1, time 2, time 3, time 4, time 5; if I go and change a working memory element, then they get a new time. The first thing a rule is picked on if a rule matches something that is most current in working memory, if that rule had a higher priority of being picked as a rule to use of the conflict set rather than the rule that matches something less current. If a rule fires in this piece of the network, the way that the system works, they don't keep on doing things around this piece of the network, but it will always be looking at the whole network. It will keep on doing things around this piece of the network until all of a sudden something matches over here that either has more knowledge or something like that. Does that answer your question?"

Ralph: "I think I see what you're doing."

D.3 The Critique

Ted: "OK, let's look at the design then and see how this turned out. I think it would be best if everyone gets on one side of the table. It's hard to read it from both sides at once."

Claud: *"Well, it doesn't matter to me if you put it upside down. I used to teach ..."*

"All right."

"I can still read it either way."

"Well, let's see if I ..."

"I haven't really thought too much about the how I do design."

"Well, that's OK."

"I guess we just do it."

"I think most people just sort of do it. Let me explain what my drafting is really all about here. These boxes are associated with ... (perhaps I should do something about ...)"

"All right."

"Do you want some tape?"

Ralph: "You may tape it down after a while."

Ted: "Let's mark the orientation."

Section D.3   The Critique   147

Claud: *"OK, might as well orient the thing."*

Ted: "It'll help later. I'll try to figure out how to put this thing back together again later. These are registers of different bit widths. That's a 64 bit register. Zero is the low order, and 63 is the high word."

"Yes."

"This is backwards. IBM does things backwards. I first learned to do a large operating system, on an IBM system. And then after six years of using DEC equipment, I finally stopped using hex and started using octal, and now the DEC community has changed it to the VAX and is back to hex, at least. The byte order is still different. The order is really screwed. It's like ah-h-h! Just take that to mean a 64 bit register. It doesn't matter which order the bits are in. So we have a number of registers. These registers here aren't connected because they're part of the channel controller. The design that I have here has floating point, it has all of the integer, packed and decimal instructions and such, and I omitted channel controller stuff because I thought that was a connection to the outside world. I don't know. In your design did you use floating point on your chip?"

"Yes. That was mostly handled in ROS. That's the reason for 17 working registers."

"OK."

"Our problem in trying to evaluate that as a true 370 is going to be a little awkward because you have to bear in mind I was doing that to also run another architecture, namely, UC1 which is a very simple machine except it's heavily register oriented, but we did those in a stack which is not shown, obviously, and so that data flow will handle either a 370. If it can handle a 370, it can handle sufficient complexity to handle any architecture I've run into so far. I've only done about three and since, I've dropped it. I haven't been working with it any more for about a year or so ago in a different area. But you will find that our busing and our registers are, in fact, more general purpose than you possibly would do. If you were really optimizing in 370 only. You implemented directly most of our architected registers, I presume?"

"Right."

"You got your gp's and you got your controls?"

Ted: "Right. The control registers are there; the general purpose register stack is there."

Claud: *"Yes."*

"So I have a whole bunch of those. My floating point registers ..."

"OK, you've got your fp's."

"Right. Let me get on the same side as you."

"Yes, get on the side with me and let's see where we are."

"I have a hard time reading upside down."

"I'm sorry for my designer who's Jerry, and he is apparently lost as usual."

"OK, so we have the storage keys ..."

"Yes, and we had those. We had ours in a stack. I can tell you where most of the stuff was."

"Why don't we write here that was in the stack."

"Yes, that was in the stack and that's due to space. Tell you why."

"All right. That's good."

"Normally I would not. Normally I would have implemented those."

"OK, let me tell you about the constraints that I used because designs change about constraints."

"What are we looking at here, 67 or is it 68?"

"Right, 68 goes to here. Constraints I used. I said, 'OK, I want this to be a fast design, 'so if there are times in which you can have multiple ALU's and increase the speed, do it. So you see, I actually have three ALU's. I also said there are some registers which I think should be architecturally defined, and those registers were like the DAT box."

"You may have noticed in our implementation, we couldn't implement the DAT box on the chip. Due to space, we have compare registers, and went back off and handle that in microcode and, again, that was due to space. We had a design with that, and it had to come out because of we couldn't wire the chip, well. No, we'd gone above my arbitrarily 5000 circuits. I laid down the law so that nobody goes above 5000 circuits."

Ted: "Why did you choose 5000 circuits?"

Claud: *"Partly because of heat constraint."*

"OK, so you had power constraint?"

"Yes, power constraint. We were in BIPOLAR."

"OK."

"And secondly, it's sort of like playing baseball where you only get one strike: you have to have hits."

"OK."

"We have the wire. There was no question, I had to wire that chip as far as I'm concerned, and I believe that I could wire 5000 circuits. And I didn't believe I could wire 7000, and we couldn't."

"OK, because your tools would blow up or ..."

"Well, no. Just not enough space. More space on that chip went to wires than went to circuits, so wire turned out to be the problem — not circuits."

"OK."

"Let's see. There were over 10,000 wires. So, it turned out to be a wire problem. So that's why in implementing our data we did the compare and as long as we stayed in the page, you'd never have go do anything more. Compares you can work with as long as you were in that page. And so we were working in addresses like that, and when you crossed the page boundaries, you'd have to do a translation and update. And that gave fairly decent performance, by the way. And those are the kind of tradeoffs in deciding what do you throw out and we threw that out and kept working registers aboard."

"OK, so we have the registers there now. I also have a primary memory."

"All right."

"And then we have this register and a buffer register — we actually have a double buffer register. I can get memory fetch ahead that way."

"OK, so you in essence cache 1 deep. We did not put a cache on ours due to cost, and that was all. It wasn't questioned. But, you see, you're building — let's see how wide your adders are. See, you got more silicon to work with. So you're building a larger machine. If I went to a larger

machine, I would use a cache, a small machine like I was working with — you see, in the range we're in, if you're working with a machine in the 8 bit to 16 bit, they're cost oriented. Because you can get a good bit of performance. And as you go into your larger machines, then performance becomes your criteria and you shift, and your machine here looks like you're in performance rather than having cost predominate. You have to be cost sensitive, but it doesn't predominate, so in this design you always have cache until memory cost pricing goes on for another five or six years, and then we may even throw it out."

Ted: "With constraints like that, the design takes too many hours to do multiple designs, when I fed in the constraints I said, well, I had a number of choices to make. Indeed, I ..."

Claud: *"Consciously made performance."*

"Consciously made performance, yes."

"All right."

"My predominant bit width through this whole thing is 3 busses: there's a 24 bit bus (which gets involved with the memory address register, and whenever it needs an address register, it seems to fall on the 24 bit bus). Most of my transfers are in 64 bits. I ended up needing the 64 bit bus, then all the 32 bit operations just fell on the bus and well, I'm just going to use only half the bus. I don't bring the wires over to the other half. And also have an 8 bit bus on which bytes are brought back and forth. What kind of busses did you have on your design?"

"All right. We had an 8 bit adder on it, and that's simply because it could keep up. I was memory bound, and that was 2 bytes wide."

"So that was 16 bits."

"Yes, that was 16 and we did have a 24 bit bus and a separate arithmetic unit for addresses. By the way, you'll discover that you do more arithmetic work calculating addresses than you do anything else."

"That's right."

"So we optimize the address calculation more than we did the other, simply to give us performance. So we had that and, let's see, what else? Predominantly we were, as I said we had a 16, 8, and 24 busses on that chip."

Ted: "A lot of my registers here are dealing with floating point, and you dealt with that on microcode in some sort."

Claud: *"Yes, we did."*

"So those things would all have gone to microcode?"

"Let me tell you, we did not optimize floating point at all. We were going to do a separate chip for floating point and did not simply because we didn't seem to find favor in the commercial side of the world with this company and so we were in the wrong division of the company. You see, there were people who had responsibility for designing computers in that performance range, and they were physically not located near where I was. So, anyway, we almost ignored floating point, so I'll put down low priority and planned a separate chip."

"So we cross off registers that were floating point."

"Now, in the speed range you're in, the chances are you would have floating point, and judging from what I'm seeing, you possibly would not optimize your floating point at the expense of data handling. Machines in this range normally are data-handling bound in performance. You'll discover that if you study even in the heavily optimized Fortran work, that you can never get that percentage of work above 30. You just can't do it."

"So your feeling then is that these registers then are justified in a more Cadillac version performance machine."

"Yes. You see, when we started out, our objective was to put a 370 data flow on one chip. It's a bit of pizzaz if you can do it, and it was just a test to find out if we could do it. So, obviously, we were constrained by the amount of hardware we had and we decided to spend it in general, optimize the arithmetic operation of address generation, and then spend the rest in trying to hold data on chip so that you don't have to clutter up the memory bus which is the bottom line on this thing. And we spent our pins primarily in getting a parallel interface on to that chip so that we could handle all of our registers, in essence as parallel as needed for control and, secondly, if we had the pins, we didn't have to decode. We saved levels which gave us performance. That's the only way we could hold 100 nanoseconds. We could access our control store and bring it back and stay within the 100 nanoseconds."

"In your read only control store, what did you have out there, and also what was it controlled for?"

Claud: *"For all registers. And the adder. It controlled the data flow and function. The functions, of course, being adders, shifters, and comparators. So it controlled not only function but the gates in order to do it and the latching. We did not have multiple aperture registers (as I call them). Where you got a stack where you can store two things in or read two at once. Now they're needed and they would possibly be needed in this for your gpr's. Oftentimes you need to read two gpr's at once, and if they're implemented in a stack ..."*

Ted: "Right now I have them as a single port memory of sorts."

"If you've got them individualized, why then fine. But if these are all implemented in what we'll call memory technology, and you can only read one of the registers per cycle, you're going to pay a penalty because normally you'll want to put a pair of those together. So if you'll look at the way 370 is done it pulls two registers and stores one so that sort of thing is very useful when you can define it and you can very easily define it that way; the question is to hand it to a technologist and tell him to build you a few."

"You brought out an interesting point about when you deal with registers, you often grab two and store one."

"In the same cycle, right."

"In these ALU's what I have often is the registers come out and go to a bus. The bus has the ALU on it. It also has a register on it — this register here. And it ended up being a general bus-stored register so the first phase of the bus would load this for one value then the other would load to another. Then they would do the ALU thing. You would actually bring dual ported registers out to your ALU's then. Is that what you're saying?"

D.4 The ICCC Design

"That's the way we handle it. Let me show you (I don't have my design with me. I should remember it by now. I've worked on it long enough.)"

"Could we draw it on the plastic here?"

"I suppose we can, but it's been a while."

"I can't take the blackboard with me. The plastic would be best."

Sumit: "Are you sure PEOPLExpress wouldn't mind the weight of the knowledge?"

Section D.4   The ICCC Design   153

Claud: *"Off hand I'll try to lay it out a little bit. Now we had an arithmetic unit over here."*

Ted: "This pen was writing fairly well for me. Would you try this one?"

"I'd as soon have one as the other. And, with that and now we had a decimal corrector immediately following it (decimal corrector/adder). Now, we come down and we had what I would call a z-bus (it goes across). There were 17 registers through here. (I'll just start putting in the registers)."

"Now, this bus was capable of going into all of those. Of course, all of these are gated and it goes into (and I'll put down 17 over here for you). Now up here, I have two busses (come across, one here and then a second one there). All of these could go to one of them. It would read 'all.' The other would read 'selected ones.' This one could read 'selected registers.'"

"How'd you choose these two busses. Did you just decided you wanted ..."

"You're always trying to get a reasonable speed for the amount of investment in hardware. All right. Since we can put a good bit of our data here in these registers that we're working with, and if this is only 1 byte wide, then we have to ..."

"How wide are the registers?"

"These are just a byte. They're eight bits, and in handling the architecture of 32 bits (it's predominant in 370); therefore, you're going to make four cycles. OK, now. In doing that, we want to be sure we can finish 32 bits in four cycles, and in so doing, that says that you will need to feed both busses each cycle. So to do that, then it says you've got to have two independent busses, you've got to be able to independently gate from two registers: one to this bus and the other one. So you have to be reasonably careful about where you put things."

"Right, right."

"And we had a set of rules about where we put things that guarantee."

"What are the rules like?"

"We held down below. We would hold our ... We could group these up to three. Registers could be grouped into three and read out on to a 24 bit bus for address generation, and that bus and an adder were 24 bits, and you could read three and normally (I believe this was an increment of some sort), an increment of $+$ or $-$ 0, 1, 2, and 3."

Ted: "Is it easier for you to design the incrementer by any constant or small constant?"

Claud: *"Small constants are much easier."*

"And what would you define as a small constant — just those 0, 1, 2, 3 ... OK."

"The next one would go to 8, obviously."

"OK."

"It's not that difficult to go on up, it's just that we couldn't find much use for it."

"And if you want to go to 8, when would you think about going to 8? If you needed it, or if there was some cost involved, or is there some sort of tradeoff to make there?"

"Yes. We would ..."

"How much more expensive is an adder if you have to do up to 8 vs. only up to 3?"

"Not very much, really. It turned out we didn't find any use for it. That's the reason we didn't do it. It's not that much more expensive. It's actually bringing in another line. Your controls are possibly more expensive than this hardware."

"OK."

"And here in three registers, we would hold the address in memory, we hold the address in our stack which was off chip, held our gpr's, well, all of our architectural defines — floating point registers and control registers ..."

"They're all off chip, OK."

"Control registers — all in a very fast stack. I could still access it. Logically, it was on chip, because I could access it in a cycle — the same with my controls. And so logically my controls down here lost ROS (read only store). It was also one cycle."

"Those were all parallel — those on two chips?"

"This is 16 bits here. It's all I could do. 16 bits and I had enough buffer in here to where I could keep going. Down here we had — what is it — 51 lines plus parity plus 3 parity on my controls. And that says I can do

all controls in parallel. I can do three things in parallel. I had some of these registers designated as IO (there was an IO bus that came in also — this was my IO bus; it was 8 bits) and you could read into several different registers, I don't know how many."

Ted: "Excuse me. Your IO is also off chip. Over here in this design, by chosing not to input the channel stuff, that actually you had off chip anyway. How did you implement your channel control your start IO instruction?"

Claud: *"That was all. This is the input bus. Actually there's another one of these chips. You can't build a controller for less than a chip."*

"Sure, sure."

"It actually was another chip like this one. I changed the ROS."

"You had to interface here your channel controller with your address registers and those things, and those were all off-chip."

"Yes."

"And you access that off-chip through with this 8 bit IO?"

"Yes. So our IO's come in, so I could handle three things at once: you could do the arithmetic, you could also your address, and you could be doing IO."

"You said address was the most predominant thing that you did. What sort of percentages would you assign to those things as to how much you were doing of each?"

"Oh, gee, I can't give you percentages. I can say that's one, that's two, that's three. [Address, arithmetic & I/O, respectively.]"

"Great. Fine. OK."

"I really don't know, but that's the priority right off ..."

"Sure."

"... with no problem. Now, as far as the toughest thing in holding the 100 nanosecond cycle turned out to be this path."

"OK, from the ROS was the toughest."

"Not only do I generate the address for main memory, I generate the address for this one."

Ted: "Oh! You did both addresses! OK, why did you choose to do that?"

Claud: *"Well, actually it's very seldom that you have to do ... excuse me. I didn't do it through here, I'm sorry. I did that with logic on chip because your address for your next order ROS word is normally sequential or it is included in a field."*

"Ah, OK."

"Then you carry it there."

"Right."

"We carry it in a field within the word; therefore, the ROS was pretty self-contained in that respect. But we could change the word based on condition of the adder. If you had an overflow, underflow, our result was zero, and this was automatically set."

"OK."

"I also had a register here which was a condition register and I had eight lines that came out that could be set from some other like my IO controller, and that was a condition, and then I had eight (I guess there were eight more or actually four) there were four that could be set from my adder."

"You said four?"

"Yes — that conditions, and that would be equals zero and overflow and I guess you'd call it underflow."

"Now, in my design I had actually a PSW word. If you took your PSW word, you spread out to the 8 bits here and then ..."

"I held part of my PSW here and the rest of it was up here. I look at this as the architecturally define."

"Right."

"Machine was here."

"Most all the architecturally defined. OK."

"That's right. That was a stack. So we did that because this was a relatively small piece. Now, this was eight bits, obviously and these, of course, are eight. There are eight bits at this point back up here. This is 24 and this, of course, comes down and is 24 and goes back in to registers. Now, so you can begin to see that wires gave me problems — as much

difficulty as anything else."

Ted: "Sure. And so the reason why you kept the eight bit data path instead of having a 32 bit data path was just cost."

Claud: *"We designed that at 16 and 32 and looked at it and we couldn't get any difference in performance."*

"Oh! OK!"

"There was no performance difference because we had relatively fast circuits. They were in the order of 4 nanoseconds, and so we even allowed our logic level depth to limit up. Normally, I was going to ask you if you could run your logic levels. Our logic levels per cycle ... Normally in logic you will usually hold that down around 14. We were up at 24 and simply because we could do 24."

"OK, I don't know what the term 'logic levels per cycle' really means. Is that the number of transitions you make per clock phase or ..."

"That's logic blocks."

"Oh, logic blocks."

"You know, like that and so on."

"OK."

"OK, I could do 24 of those in a 4 nanosecond cycle and so we did. Actually it was 3.6 but that would only be 96 nanoseconds and I was running at a 100 nanosecond cycle and the statistical distribution that's worse case for 96 — statistical distribution. Since that's worse case, you're way back over there. It actually runs about 70."

"OK, I don't have a metric for logic level per cycle. You see, each one of these modules I have defined as to how much delay each block takes."

"That's good enough."

"Then I go and I add the number of blocks I had per clock ... AND'ing and OR'ing that stuff would all just go blah, blah, blah in one clock phase."

"OK."

"And it keeps track of how much it needs before it has to go to the next clock."

Claud: *"As long as you know the delay through a function, that's sufficient because that's how I use the number level just to generate delay and that's all."*

Ted: "Right. In fact one of the things that gave me a lot of grief actually in doing these rules is designers kept on saying, 'Every time you're done with a clock, you'd better have everything in registers or else I'm going to scream at you because I can't use LSSD techniques, I can't use testing techniques.' So some of my registers are there because I had to go and latch values up to clock phase."

"By the way, I couldn't use LSSD on this because I didn't have enough blocks, and so what I did was I set up where I had a test port (a test port I call it), and that was so that I could put data in and set any register on that chip by my test port. And I would set them as a byte at a time, and I'd actually use one of the ports, but I had to have some micro code in order to do that. And I could go into test mode, and I could load any register broadside and then read it out. Therefore, I had no logic on here that I couldn't get to with the exception of implementing some of the console function that's architecturally defined, some pushbutton stuff. Because I don't know how you test ... When you push button and get only one pulse out."

"That's right."

"It's a real dog! In fact, I don't know how you can do that with DC tests."

"I don't know either."

"That was a piece of logic that we couldn't check."

"Sure."

"The rest of it, why we checked it that way. We used a test port rather than LSSD because we couldn't get in but we used the same patterns."

"Do you remember how many words you had in ROS?"

"Yes, I think I do. I believe it was 2000."

"2000?"

"Yes, 2K words."

"OK."

Claud: *"It's two or three because it comes in 2K chunks. That's the reason I know. If it had come in three, we would have used three like on the Mod 50 which was my machine — Model 50, 360 machine, while we used 2800 (2.8K) words which were a little wider than that ... words of ROS. Now if you look at that and say it's a waste of pins to go 51 wide on your control (and it would be if you were designing specifically for 370) and you can narrow that down. I didn't know what other architectures I was working with. This was an experimental chip, so you hold your options open. This gives me more variability, and so I did it that way. The reason it would be in real life you'd squeeze that down. Pins cost money one dollar per chip. They're expensive so ... A buck changes, of course. What I'm trying to say is ..."*

Ted: "I understand."

"They add significantly to the cost of your chip and to your next thing you mount your chip on, and then it goes on and on and that's why you'll notice we're using more pins than anybody else. But the reason they don't use them may not be that they don't know how to put more pins on the chip."

"The cost."

"It's that cost squeeze. They want to keep the cost down."

"Actually I've screwed around with using my system to generate some other designs and, you know, one of the things my designers was telling me is that a system like this is wonderful because most times he'll find he's got area to spare on his chip because it's pin out limited. He can only put so much function on his chip because of his pins."

"OK."

"And so he doesn't care if he wastes a little area here or there as long as he can get fast turnaround in getting that architecture from the algorithm."

"Well, I was also pushing the number of pins. I was using more pins than we do on our commercial chip."

"How many pins did you use on the ICCC chip?"

"200."

"200 pins. OK."

Claud: *"My original design was 240. They got nervous, so I cut it back to 200."*

Ted: "OK."

"We can make a 240 pin chip."

"Something I've been working most recently on in terms of my rules in choosing busses: how do you choose how to put things on busses and not put things on busses."

"OK."

"How do you choose major bus widths? How do you do that?"

"That's due to the exclusivity of your data, like here. I decided I needed two inputs so I have to have two independent busses. Now the way that I would tell my compiler to handle temporary data is that you've got to look ahead and see what you're going to be doing in the next cycle and be sure that they can feed both busses. And that's the way I set up the gating."

"OK."

"And it was that set of rules that guided us in coding our ROS to decide, OK, where are you going to put your temporary data."

D.4.1 Tape change.

"We started out, I had these constraints: 5000 logic blocks we had, I'll say 200 pins, and I had three watts. All right? That was what I had."

"Now how did you, OK, can you tell me how you went from watts and pins to blocks?"

"OK."

"How do they relate?"

"All right. Blocks come in two flavors: there's fast blocks that are hot."

"Sure."

"And then there's the slow ones."

"Sure."

"Slower. And when you need a fast one, or the drivers are hot. The drivers that drive the 200 pins, represented almost half the power of my chip."

Ted: "OK, sure."

Claud: *"OK, so that's the first thing you look at. You say, well out of the 200 pins, how many are drivers? Rule of thumb is 40%."*

Sumit: "Then that's two to one then?"

"Not quite."

Sumit: "Two out of five. Right?"

"Yeah. Not quite. A conservative guess is about 40%. And so when you first start out, you know, with a blank sheet of paper, that these are the kind of rules of thumb that you start with. And you start there and you say, well, oh, roughly 40% will be drivers. Some of them you'll have both drivers and receivers on. We had that capability of handling both drivers and receivers. Yeah, we could only have, I don't remember how many drivers we could have; it seemed to me like 140, something like that. Now that is open to question of my memory, but we had a lot of receivers. Anyway ..."

Sumit: "Wasn't there, excuse me. Wasn't there also a limitation on the number of simultaneous switching drivers?"

"Yes."

Sumit: "Yes. And 40% still put you way below that limit, right?"

"Yeah. No, no, not below the problem. No."

Sumit: "It would still be way over that limit."

"Yeah. But never the less, you start out, you subtract that power and decide what you've got left for logic."

Ted: "OK."

"And then you immediately determine, all right, I can use 10% or 15% of the high power blocks."

"OK."

"Whatever percentage that is, and I don't remember what ours was, but I imagine that it would be in the 15% to 20% range."

"And then you use those type of things, you use those for buses?"

"You'll use those in driving this bus ..."

Ted: "OK."

Claud: "... *that bus or that bus.*"

"As a bus driver. Right."

"*Yes sir. Because this line winds out being relatively heavy capacitance load.*"

"Sure."

"*And we took it and broke it up. That bus I drew as if it was one line. In reality it's not. It turns out to be the one that has the most inputs on it was four. So you had to take that bus and make it a Christmas tree to get in there. And this one was two. A two-way bus the other one was, so, I mean it's a two-way tree to get in. And in doing that way you use power, and then of course you use, I believe these are two-way trees, I'm not sure about those. But, nevertheless, it's those kind of things that begin to chew up your power.*"

"Mmm-huh."

"*Then you do an estimate on your adder, which is relatively small, by the way.*"

"Sure."

"*That's trivial.*"

"Your ALU did what sort of functions? Did it compare, add, and subtract?"

"*Yeah.*"

"Did it do any shifting?"

"*We did shifting outside the ALU. Not in that, but down here in our corrector before we did that. But, it would also do the logic the and/or, exclusive or, of course and invert.*"

"Sure."

"*And, over here, we didn't have shifting.*"

"Here's some places where it had 8-bits of anding. You actually would have folded that in the ALU?"

"*Yeah, that is in the ALU, yeah. It's cheaper and fundamentally, we didn't want to have to load that bus any more.*"

Ted: "OK."

Claud: *"It would possibly not have cost any more to have done that, but it's actually cheaper to do it this way. It's just that once you've got it to where it'll add and subtract, or add and compliment/add, it's very simple to get the rest of it."*

"That's an interesting difference between INTEL, this is, actually I added, before I used to go and fold the stuff in. A designer that we work with at INTEL and said, 'Take it out.'"

"Take it out, I keep it out."

"It's just stylistically the way you do it."

"No, it's easier to debug."

"Ah! It's easier to debug! That's why he does it."

"Yes, sir, oh yeah. Yeah, it's easier to debug. And I sympathize with him."

"OK, so that's why he does it. Interesting."

Ralph: "How much, you know, it may be a general question not related to exactly this, how much is, you know, when you bring up INTEL, I think there's another difference in that we're in the BIPOLAR. We are heavily into the gate array, and they're in the MOS design."

"Oh, yes, by the way, this is gate array."

Ted: "Sure, OK."

Ralph: "Obviously ..."

"So this is a gate array design. So what you have from INTEL is basically a MOS design?"

Ted: "In my thesis I'm going to be talking about this and where I find differences between the two, you know, all I can see, actually, in an issue about knowledge acquisition, multiple expert knowledge acquisition is a real hose, I mean because experts have differences of opinion. And that's good. They're supposed to."

"Sure."

Sumit: "They have their biases."

Claud: *"Well?"*

Sumit: "And their value systems."

Ted: "Sure. And one of the parts of things that people keep on trying to do is say well how do you get multiple experts to agree on something. I think that's totally wrong! You don't want them to agree."

Sumit: "No, no."

Ted: "Because they're different. They're supposed to keep their different things in so you can have different rules for ..."

"... different rules for different parts of the thing."

Sumit: "Do you also, shouldn't you also have the similar assumptions?"

Ted: "Sure."

Sumit: "Why don't you try to combine them together to have a set of assumptions, and maybe once you understand the assumptions, then maybe you can find some similarity ..."

Ted: "Right, right."

Sumit: "... you know, between their value systems."

"Well, that's, yeah. I design differently depending on what I perceive to be the criterion that I'm trying to satisfy. Like this was when I first forayed into the smaller stuff since, oh, 25 years."

Ted: "What have you done over the 25 years? You mentioned ..."

"I worked on the 701 which was the first one we ever built."

"Really, I've got a plaque on my desk of a little, you know, the plaque of the little plate from the 701. Yeah."

"I came up here in 1951 or 1952 or so."

"Wow, that's a really neat machine. We had that at the University of Michigan when I first got there."

"Yeah, well this would roughly have been, what, ten times or so or more? Not quite."

"So you worked on the 701?"

"The 701, the 702. I worked on all the 700 series and taught 701 and 702. And, let's see, the 7074 was the first one that I managed."

Ted: "Uh-huh."

Claud: *"And I did the model 50. Actually I had the channels for the model 50, but the manager was never in town, so I did his."*

"I've done that before."

"And I did FAA ..."

"Uh-huh."

"... the FAA system — you may not be familiar with that one."

"I've never heard of that one before."

"Perhaps, let's see. That's a 7-way multiprocessor into 12 memory boxes with 160 gp lines."

"Wow! You've designed a lot of control."

"I did that 20 years ago almost, and it's still in use, by the way. Any time you fly, you're in my system if you're in the continental United States."

"Really! Neat!"

"Or in London."

"Uh-huh."

"If you go into the field of air. That was the most fun out of anything I ever did. And since then, oh, I did the chip and I've done some architecture. I did the architecture for the Model 67, part of, I can't take credit for all that. I did the relocate. So I've worked with that sort of stuff. It depends on what it is you're trying to do. This case, I was trying to push gate array. My goal was to find if our gate array tools would handle 5000 circuits, and the vehicle I chose was a 370 Data path."

"Sure, sure."

"OK, now, that was what I really set out to do, was to see if our tools would push on out to there and turns out they will."

"Uh-huh."

"With a little bit of changes made."

"Well, there's always debugging."

Claud: *"Yeah."*

Ted: "When I first started running the 370 through, I had run 6502 designs, PDP-8 designs. I thought, hey those rules are solid."

"Right."

"No problem."

"That's right."

"It's about a third of the way through and all of a sudden, it goes into an unusual loop."

"Excuse me?"

"I had never seen this before."

"Yep."

"Uh-huh. And while I did a little bit of quick debugging and right, uh-huh."

"Yeah."

"Tools are interesting when you push them ..."

"Yeah."

"... you really push them."

"Well, we blew most of the tables because we had more data than they were used to seeing."

"This was a name stack table that blew up."

"Yeah. And so, with the help of the EDS people, we modified, and got it going. They finally wired it. Now what else can I help you with here?"

Sumit: "Would this be a convenient time for a coffee break?"

"Right now!"

Ted: "That sounds good."

D.4.2 Coffee break.

D.5 Tools

"I think about what I would look for if I were coming to you and say, 'I think that we can build tools that will greatly assist design.' Let me tell you, as a designer, what I look for."

Ted: "Sure."

Claud: *"Now in those I can put up, and that possibly is for future research, not for you. You need to write this up so you can get a thesis out of it, I know it. That is what's required here. It appears within the constraints that you have on it, to be doing the job."*

"Right."

"Now I operate under slightly different ones, but it's a possibility it's the next level down."

"Yeah, yeah, sure."

"So it may be an extension to this tool, maybe a different one. Don't know, and I'll show you what they are because I see it sort of sitting on three things: one of them is that you have to be able to in essence, implement an architecture, OK, in functional units."

"Right."

"Whatever we call them."

"Sure."

"All right. When you do that, then the next step will be, all right, let's take and map that into the physical."

"Sure."

"And that has a totally different set of constraints."

"Sure."

"Depending on whether it's MOS or it's BIPOLAR, and the size of the chip, the amount of logic you can put on it, and that sort of stuff. So that is, I would say, another extension and we impose still another on ourselves and that is that we be able to test that it was manufactured correctly."

"LSSD."

"And testing winds up to being a very interesting field all by itself."

"That's right."

"You test for three different things. First, you test the logic designer, to find out did he do his correctly."

Ted: "That's right."

Claud: *"Will it do the function that he thought it did?"*

"So up here like, up here in my architecture ..."

"And that's a verification."

"When I'm sitting and chosing this architecture, I have made sure that I can get at clock transitions, everything is in registers. We can use LSSD's, that was not easy, that was a lot of work."

"Right."

"That's something that's important at the time level."

"Yet functionally, you see, unless your advisor understands that, why he could look at it and say, 'Why was that registered? What was bothering you?'"

"Yeah, and so, we have to test for the accuracy of the design. Number 1, number 2, manufacturing wants to see if they can replicate it."

"Uh-huh."

"... by the thousand, and then, Number 3, the other guy we test for is the service guy, who just wants to identify the one that's bad, so he can pull it out. And those are three different kinds of patterns I mean ..."

"Sure."

"... and yet you've got to do them all."

Sumit: "However, in some designs, though, that the service guys are beginning to finally wake up to the advantages of LSSD and more and more they are using it, you know, your 3081 series of machines uses it."

"Well, LSSD is a technique. What you really want to do is be able to, well ..."

Ted: "To use the tools."

"... to do a functional test that will identify where you are malfunctioning. And LSSD is a very good technique for doing that."

Sumit: "That's right. Because it's like a LSI scope."

Ted: "Yeah, it's a method of loading a pattern."

Sumit: "Yes, exactly."

Claud: *"That's the way I look at it. I don't really worry about it. And the fact that we have to invest a million dollars into the software, makes it very ..."*

Ted: "Yes."

"... easy for us to use."

"Well, to further answer your question, since you seem to feel that this design is given the constraints that I said, this system should do and what it produced, and that you feel happy with. I plan on taking this thing and writing this up, and then I've got to do an analysis of my rules, what's actually in them, and how my system ran — the normal set of analysis you do in a thesis, but I think a couple month's worth of writing, and I hope to be done by Octoberish."

"OK, now I don't know. I think it would be useful if you have some 370 data flow, you know, big gross things for him, like a, oh, a 138-48, or even a 58; a 158 would be a good number, a good low number. Something like that for you to look at a little bit."

Sumit: "OK, I don't have it with me."

"Well, however, I believe that if you can take those ..."

Ted: "Uh-huh."

"... since you know exactly what you have here ..."

"Right."

"... and then you look at that, in a moment you can do the translations."

"... Right."

"Say, I've got this and not that, and so on."

"Right."

"Now you may need to talk to somebody to understand why it was done the way it was, but you have to bear in mind that we often times will leave something off simply because it physically won't fit on the board."

"Sure."

"OK, we say, 'That's all the space we've got and we're just not going to buy another board or panel because sometimes that means another

frame.'"

Ted: "Right."

Claud: *"And it just says, 'Hey, you're out of the ball game, get back in that frame.'"*

"I take it it's a business decision."

"Well, decisions, you see ..."

"... Are made on many levels. Yes, I understand."

"Yeah."

"Well, the reason why I want to take this and compare it to your design is because I understand your decisions. OK?"

"OK."

"And then I'll look and I'll say, 'Well look, I have my registers here or up there on that stack ...'"

"Right."

'... I have data passed down to bigger because you wanted those performances.'"

"I want to go see if I've got any data if you'll take his address so I can see if ... Tom ... Ted ... No, no. OK, and I can send you whatever data I've got."

"That would be wonderful, I'd really appreciate that."

"It's useful because it ... I hesitate to hold that up as 'the' design ..."

"Well ..."

"... it is a design."

"No, I understand. Claude, you see, that's all I care about."

"OK, fine."

"I would appreciate any more information you have about, you know, as a consultant."

"Yeah, cause I don't consider it, yeah. I'm about to clean out my files so I can ..."

Section D.5 Tools

Ted: "A four-page ICCC paper was pretty sparse. I did a 6502 on a four-page DA paper. Boy, you had to really read between the lines to understand anything, you know. Three sentences — that was all."

Claud: *"That paper, at the time it went out, was pretty heavily edited, I'd say."*

"Oh, sure."

"That paper almost didn't go out."

Sumit: "You know Don Thomas has a research contract from our department here to look at knowledge based synthesis. Is that eh, you know I suppose ..."

Ted: "Yeah, that's my work."

Sumit: "Yeah, that's your work."

"That was my understanding. That was the reason I thought we ought to give him a little bit of help here."

Ted: "Yeah, I'd like to talk about the tools, too, more."

"When I look at tools what I want is first and foremost, I want to be a record keeper. I don't expect tools to design for me, let me say and adder. Because all the adders I have ever needed have been designed."

"I agree."

"You go to the library and look up people with theses and all that, and you just don't need another one. So, I would expect them to handle an adder simply as a function box with inputs and the timing."

"Right."

"And that's all I want to see ..."

"Right."

"With a delay, all right? Now I'll need a method. When I get to it, to technology dependent stuff like the amount of power and delays. So you need a technique let it be general or fill in details that are specific to a technology. But, I want to be able to have functions such as: adders, ands, ors, exclusive ors, and multiplexors, registers. We did this design using, oh, a language you've probably never heard of called BOTUC, which is a private one."

Ted: "Sure."

Claud: *"I used it for a very strong technical reason that it was the only language used in the German lab at the time I happened to be in Germany when I did this, so I was over there for some work, so I used their language."*

"Yeah. But BOTUC is a register transfer-level language or a like do you describe the actual registers and how they have to get things around, or do describe something that you want to have optimized?"

"No, it's a register entry. I do the optimizing."

"Right, OK."

"It's a record keeper, fundamentally."

"OK, yeah."

"But it does draw all my blocks for me and all that kind of stuff. It is very handy in that respect, and I can make changes and it will update wherever it goes. It's a nice, what I call, logic entry language and diagram drawing for me. So, I've wanted to handle registers where I define the width and the controls that I need. Now, and I'd want it to be able to handle parity generators and all that kind of stuff. So, I'd want it to handle blocks at that level. Now, at the time I move toward technology, I want it to keep track of my pins, the amount of power, without me having to write it down. I'll give him the rules, and any time, the fact that I'd used 478 blocks of this mixture. I'd like to ask write power, get power, space, get space, and that space for the active stuff, if I'm working at the chip level. Now, most people in IBM now, are becoming specialists. All large companies, in fact I had a talk Monday morning on this, but we originally start as craftsmen, where we do everything, and I've gone through that stage a long time ago, and then you gradually start using the tools that are around and eventually you become expert with the tools."

"Sure."

"And you have lost the big picture."

"Right?"

"And somebody comes rushing in and says, 'Hey, will you, here's the logic I'd like to have all the test patterns in order to test it.' And you take that to the testing expert."

Ted: "OK."

Claud: *"And so we gradually become groups of experts, and so I find that when we do that, then we need tools that talk to those experts. And there's a few of us around who flit from expert to expert and try to talk all the languages and have difficulty with all of them. Yeah, and so you'll be, yours is specifically for a logic designer who has had a piece of the system cut out ..."*

"Mm-huh."

"... and they'll say, 'You're going to design for me, well in this case it happens to be the arithmetic part of our system."

"Right."

"We've got another kind there's the channel ..."

"Right."

"... OK? and so they'll need to be able to describe functionally what they want as input and I want that to keep track of all the details, and, you want the interface defined to this larger block, they call it hierarchy, perhaps."

"How it relates to your disk drives and the like?"

"Right, OK and, if you want to look at it, that's where you started with the 370 architecture. That, if you think of that as a block, that's the interface to the programmer."

"Mm-hum, sure. Yeah, in case your curious, this is actually the eh ..."

"So, everybody's got another block that's bigger than the one he's working at."

D.6 The DAA System

"This is the actual input descriptions for my system, the ISPS description where it actually describes."

"OK, all right."

"This is what my system starts from and actually generates this output here."

"OK, byte memory, word memory. OK, I clearly don't understand this, but I haven't studied your language. Maybe I don't need to."

Ted: "In general, OK, it sort of says, OK, we have a byte memory."

Claud: *"Yeah."*

"It goes from zero to some ..."

"Number of bits."

"Number of bits and it's 8 bits wide." OK, then they have mapping so you can use it as a half word or a full word or a double work in your operations, so you can just use that and it will automatically do the mapping for you."

"Fine."

"Some registers are: data register, the memory/address register, buffer register, double buffer register."

"OK."

"So you have those declarations."

"That's the kind of thing I would want to start with."

"OK, and so you say registers, control registers, floating point registers and you have the convenience of either dealing with a floating point register or a double floating point register."

"Yeah, all right."

"Going through all your registers ..."

"So you have defined the, you've put in the architectural registers. OK, all the architectural defined registers you have."

"Right."

"Isn't this specifically for the IBM?"

"Yeah, this is the IBM 370."

"You ... this has been bothering me; do you have software for each machine that you try to simulate?"

"Eh, my system is the same for all the systems but the ISPS description, the input to my system is different for each one of the things. So I have a ten-page description of the 6502, an eight-page description of the PDP-8."

"This is like a set of rules that you are defining?"

Section D.6 The DAA System

Ted: "No, this isn't the rules, this is the behavior."

Claud: *"Yeah, the behavior."*

"I have a 270-page description of the VAX because the VAX includes a lot of addressing modes, so those are the behaviors I have to implement, and my rules stay constant over all of them."

"OK, that's right."

"By the way, have you done a talk with, well I'm sure you have, with Newell and all that crew there, but at DEC, that you should have fairly interesting insight as to why they want to do busing and why they had to leave it in their systems. For flexibility, and they left it for speed."

"Yeah."

"They went to the SBI, and then there is an interesting thing on the VAX in that you look at the 780 and then going to the 750 they had just a UNIBUS first then they had a mass-bus later ..."

"Yeah."

"... for cost reasons. The 730 doesn't have really a UNIBUS either, it just has an internal bus."

"OK, before we leave it I want to make a point. They found the same thing that I found."

"That's right."

"And that is that you just can't work on that data if it's not there yet."

"That's right. You got to get it there."

"And looking at theirs I believe that they would have a good time pushing the speed up."

"... as something that checks your keys, OK, here's memory management."

"OK."

"Let's check your key."

"You've gotta do that."

"Am I doing that?"

"OK, now you're down to small functions."

Ted: "Right."

Claud: *"Let's check the keys which in reality says go read this and bring it in to somewhere and compare and ..."*

"See, if I'm running privileged."

"OK."

"If I'm not running privileged, it aborts."

"So test constraints first."

"Right. So check privilege, and then check a couple of registers to see if the keys are right, if they are then great, go ahead."

"Yeah, OK."

"If you look at individual ... we have a bunch of those service routines for reading words."

"OK."

"Reading bytes, read a half-word means we have to decode the memory/address register, go in to the DAT box with the memory/address register, come up with the memory/buffer register from the half-word memory and after we've gone and done our address translation and get our underlying thing from memory two words long."

"OK, OK."

"So, you define things as these little tiny modules for each one of these service, and then have a huge bunch of decodes for instructions."

"Oh, yes."

"The decodes for instructions are crazy. On OP code, WHOP!, you know, or on the other OP codes ..."

"Yeah."

"... then each one of the instructions."

"Yeah. Then you got to take those individually ..."

"Load-negative long, so that first checks the flowing point register two. Then you know, so ..."

"The normal dum, dum, dum, dum, dum."

Section D.6   The DAA System   177

Ted: "Right. That's what I start with. That's what I started with when I did this."

Claud: "OK, now how does it use this? Let's say that you suddenly discover that hey, I've just come upon this load and test long right here."

"OK."

"What does your program do?"

"Sure."

"It came to that."

"It comes to that, and this actually gets expanded to, this is looked at and I look at each one of the blocks like this has to call that routine. So I have to make sure ..."

"The one labeled begin?"

"Well, begin/end is just the delimiters of this routine."

"Oh."

"This check floating point registers, two. So I have to make sure that all the inputs to that are in their stable registers. So I go through and check all the registers and I make sure they're stable. Then, next, I go through and I make sure that there is the data path from FR-2 to FR-1. FR is the floating point register; it's indexed by register 2, so I have to get register 2. To get register 2, I have set up a memory, so that's in the memory register file."

"Yeah."

"So I know I have to set up the memory/address register for the register file, get its output, its output goes into the memory/address register for the floating point register file, so I have to make sure that data path there. Then, the output from that goes in ... then I set up these things so I'm ready to address the right place and finally, shove the value over."

"So I assign clock cycles cases for all those things. I assign the data paths for all those things and, in this case, I had no hardware operators, so there's nothing really much to do there."

"Yeah."

"Let's go to someplace that does like ... OK, here we had compared half-words."

Ted: "Yeah, compare half-words." "We read a half-word so I make sure all those values are available. I do a two's compliment assignment into the memory buffer register from the half-word of the buffer register memory. So, a data path has to exist that I can ..."

Claud: *"Yeah."*

"... A two's compliment move means I kept those bits and do a sign extension so I have to have a sign-extender someplace or the ALU has to do that for me. So I take the data path through the ALU and back."

"Um-hum."

"This register has to go through an ALU that does subtraction from that, so I have to have data paths for that."

"Yeah."

"And then I get to the subtraction and then I say, 'Ah, subtraction, well, how much does it cost. OK, what other modules do I have around? Do I have an ALU that does subtraction? Is it the proper bit width? If I don't, what do I have around?' Let's say I have an ALU that does add."

"Um-hum."

"Well, subtract means I'd need a role of inverters and an increment over and above the addition. So that costs something. If I were doing an addition, so I look at my cost for things; based on the cost, I say, 'Well, is it worth doing it?' Based on the cost and based on partitioning information, I divide my picture of the world up into pieces. And I know how much this module does the same sort of thing that this module does, so how close are those modules?"

"Yeah."

"So, I get a guess at whether they're going to lay out in the same part of the chip."

"Mmm-huh."

"So it may be that I have an adder over here but I really don't want to combine, because the adder over here, the data paths are all local here, and the data paths are local there."

"Yeah."

"And you know, the capacitance on a line cost you everything."

Claud: *"Right."*

Ted: "So I use all that information, and then I either combined that with something that's already there, or I make a new one."

"Mmm-huh."

"And, I of course, I either combine it with something there, or I say, well, can I make another one? Um, if I can't, am I allowed to make another one? Am I going to violate the size constraint? If I'm going to violate the size constraint, then, what I have to do is, [change] my clock cycles so I can make this fit someplace else."

"Mmm-huh, OK."

"Someplace where it will fit. So I'd use changing clock cycles to make it fit changing, you know ..."

"Can you have a technique of swapping space for time and vice versa?"

"OK."

"Yeah, that was what I was looking for."

"That's been ..."

"Well, designers do that all the time."

"Yeah, you have to do that. If you don't do that, the system wouldn't be worth the salt."

"Right."

"Um, and so that's the sort of thing, then whatever else. A zero. So I have to load a constant into there, and then I do some stuff up to the controller and back."

"To grossly over-simplify, you first test to see if you have the functional capability."

"Right."

"Of executing the function that is defined by that particular whatever it turned out to be ..."

"Yeah."

"... compared. And if you don't, then you consider generating whatever it is you need ..."

Ted: "Right."

Claud: *"You add in a register, or a path, or a gate or something, and then you assign your clockings."*

"That's right."

"I've first assigned clockings and I may re-map my clockings depending upon whether I violate the constraint or not."

"Oh, all right."

"I start, and I go through, and assign them all to start with and I may end up bumping them later."

"Yeah, well we mentally, I mean, I hadn't thought about trying to write down the general rules about how you design instructions, but fundamentally, we go through roughly that same thing."

"Sure."

"And, often times, we assume we've got the registers and all, we don't really. I mean, lay out enough of 'em so we've got enough registers and just go ahead."

"Sure."

"We just decide to assign."

"Well, you see, this goes back to what you mentioned earlier. Since this is a system, it goes and does the detail for you."

"Yeah, yeah."

"So ..."

"Well, that'll be helpful, because that gets to be, uh, it's a routine, detail that's boring."

"That's right."

"And most of us would rather not do that."

"Yeah. Eventually after we get through the preliminary stages, which may last five years, I see us being able to not only do this, but to put in; I can optimize, to. OK, you can call it speed, or you can call it ..."

"Sure."

Claud: *"... uh, lack of delay. I'm trying to think of the word for less delay, or power or space. And I see us being able to say, 'Let's look at that design optimized, to, and you'll put down.' Because mentally, you're doing one or the other all the while."*

Ted: "Right. And sometimes you ..."

"And you may put it and say, 'Designed within these constraints that the maximum delay is ...'"

"Right."

"The maximum power is, the maximum area is, and say, 'Can it be designed?'"

"In fact, I've seen experimental programs run against those constraints ..."

"Right."

"... and produce results."

"Right."

"But, it was run against history tables as opposed to anything close to artificial intelligence."

"But, there are several places in my system where, if to make one constraint would violate another, you know, it too will say, 'Look, you just can't do this design."

"Yeah."

"And because, you know, I have the constraints in for size, I have the constraints in for speed, I don't have anything for power."

"You see, that turns out to be, uh, we can translate that from logic blocks ..."

"Yeah."

"... which can be done from the function blocks that go in. That happens to be technology dependent."

"Yeah."

"As long as they can be put in technology dependent, why then that's all right."

"Yeah."

Claud: *"But, I think that the ones that are general you should be able to handle ..."*

Ted: "Sure. There's no reason why not to put them in later."

"It's the technology dependent ones that we'll add when we put in the blocks."

Sumit: "But Ted, I've got a question."

Ted: "Uh-huh."

Sumit: "Some time back you were talking about, you know, essentially the 370 design. You were talking about clocking, etc. Uh, what about those situations where there's an asynchronous boundary between two units. Do you handle asynchronous interfaces?"

Ted: "Yeah, well, what I do is, uhm, operations can extend over one clock phase."

Sumit: "Umm-hum."

Ted: "OK, but an operation can extend over a clock phase, but boundaries between operations, I don't allow to extend over a clock phase. What I mean by that is, let's say — here's my clock going on, all right."

"Um-hum."

"... if a multiply takes, let's say we're optimistic, let's say it takes two clock phases, OK, uhm, let's say it takes two clock phases and a little bit."

"And a little bit more."

"... OK, uhm, and I want to do an AND of that output, OK, I will let the AND go here."

Sumit: "Uhm-hum."

Ted: "Let's say I want to do an AND of something else, which would happen to lie over here."

"Uhm-hum."

"At this point in time, I make sure that the stuff is latched in the register. OK, it doesn't matter to me whether this extends however long; it's just that whenever I have a choice between this going to the next transition, if this is in a clock phase it's fine; if it's going to go over a clock phase, then I put it a register down."

Claud: *"What about the hand-shaking stuff, where you don't have the luxury of sending data back and forth in a channel, for instance?"*

Ted: "It's totally synchronous. It's synchronous design, CPU designs, yeah, and one of the things that you have to be very careful about in thinking about doing expert systems in your department is that three things: 1) Is that what you're trying to do an expert system for, you have an expert close at hand ..."

"Yes."

"OK, that you can deal with a lot. 2) So you have to have the person who is an expert in it and can do something. The second thing you need is, you have to pick a small domain. OK, you can't pick the whole world to start with. OK, the third thing is you have to get a prototype system working quickly. Asking someone like Claude how to do a design is a lose, all right, because Claude is very good at looking at this and saying, 'Oh, gee, I had that stuff over in this, and I had this over there.' You have to give him something to start with. He's good to look at something and tell you what to do from there, but he can critique; he's a good critiquer, but in generalities. We could talk for hours and we'd find out about his history and the war stories, but you're not going to find out about, other than a few little points of knowledge, here and there, you're not going to find out about the whole thing. So, you've got, you know, small domain, a prototype system, debugging, debugging like crazy, and an expert available. You need those three things."

D.7 Back To Critique

"By the way, while we're on that, if you, uh, find things that detract from your performance ..."

"Uh-huh."

"... often times in looking at machines that were not designed for 370, trying to simulate 370, it is the condition codes that really nail them to the wall."

"Why is that?"

Well, they're a bit awkward in 370 compared to some of the others, apparently and they can't get the bits over where they want them, and so on, and they waste an awful lot of time, of shifting them about. So, if you, you might look at a VAX; I've never looked at a VAX, interpreting a 370 and see what the condition codes and that sort of stuff is a pain in the neck for people to handle. And so, in designing these, when you get down to the hardware part, you always hold condition codes where they're readily available, because you can't afford the cost of those extra cycles."

Ted: "Yeah, that's actually something very interesting, or something I want to ask. When you have something like a PSW which has a whole bunch of junk ..."

Claud: *"Yeah."*

"... it really, is in a sense, is a whole bunch of separate registers; one-bit registers."

"It is."

"And, uh, when you're dealing with a design, do you actually think of it as separate one-bit registers?"

"Yeah."

"Yeah, cause ..."

"We conceptually say, PSW and short as a register only so that we know we're talking about a system status ..."

"Yeah."

"... fundamentally."

"You may be amused to see, let's see, where's the, uh ..."

"But, it's a lot of unrelated data ..."

"Yeah."

"... is what it boils down to. It's only related in the context of defining the system in a period of time as far as the rest of it being related, it's not."

"This is the real output from my system rather than this drawing ..."

"All right, you may have to show us how to read what you've got."

"This is actually the PSW module, I know cause that's register 14."

"OK."

"You can see that it's, you know, uh, different fields of it are getting loaded. All kinds of ..."

Section D.7 Back To Critique

Claud: *"Yeah."*

Ted: *"... here's an input, we're getting fields from all over the place."*

"Mmmm."

"Different ones like here's a couple bits here going to bits there, here's uh, you know, 15, 16 bits there, 7 bits, there's 18, you know, and the output of this mess, look at this. Good Lord! There's two bits go there, there's the instruction register product stuff ..."

"Uh-huh."

"... that goes out there. There's one bit there; there's another bit here."

"Yeah."

"There's another, you know, so ..."

"No, it's really ..."

"In terms of how to read this, module name is a register uh ... Register 14. Right."

"A port it's an output port; it's 64 bits wide and that port has different connections to it, of bit-width connection 55 to 52, goes to this other input mux and it goes to that specific port ... So you can see that those four bits go to those. And then down here, I have references back to my original ISPS."

"The PSW is uh ..."

"That's a real ..."

"... Yeah, it gathers data from everywhere."

"Then the other output ... This here would be compiled into your stuff for ROS; this is symbolic ROS in a sense."

"Oh, OK."

"Uh, for control step two to two, on this particular VT body, I don't know what happened to the top of this printout; I quickly grabbed this up; these lines have to be active, so the FWRITE from there, and those mux ports have to be active. So, I know what ports and what lines have to be active for each one of my control steps."

"OK, then you know what bits to put on."

Ted: "Sure."

Claud: *"You can take this and translate that so that it will actually cut ROS for you?"*

"Yeah, there's a guy by the name of Nagle who has a program that takes similar ROS stuff and not only compiles it into ROS, but optimally chooses the width of the ROS, and the depth of the ROS."

"Oh."

"... so that's kind of cool."

"Um-hum."

"That's what he did for his Ph.D, and there is someone who is now taking that stuff and extending it so it can have multiple controllers."

"Yeah."

"Uhm, which you know, I don't know how to do that. Most microprocessors have one controller and that's hard enough ..."

D.7.1 Tape change.

"That takes an awful lot of technique."

"Sure."

"And, uh, I guess I hold the patent on one that we used in 7074 back in '59."

"Uh-hum."

"We did, but that was a decimal machine, which I don't, well, we did a 2 out of 5 code. I don't understand why we did that. Well I know why I did; it's because I was designing a machine that was compatible with another machine that was 2 out of 5 code that was generated by someone ... who didn't have a full understanding of coding. But, nevertheless, I would do a multiply in a machine of this power; I would actually do multiply rather than the add/shift. Add/shift is not bad, but it doesn't cost that much to make a multiplier. Now, this one, you say, 'Controller — zero.'"

"Those are my inputs, to, like you had your ROS, and you had some inputs you had to bring over. OK, those are the inputs that I bring over, to my controller so that we can look at and make certain decisions about ..."

Claud: *"Oh."*

Ted: *"... what goes."*

"Oh, OK, all right. Then these, in essence, are modifying conditions ..."

"Right."

"... for my control."

"Like you can see that this, AND of a couple of lines that came over go into the controller."

"OK, uhm."

"Let's see, where do those lines come from? Those lines come from probably the instruction register. Yeah, those, oh, those are coming from the memory; wait, let me follow it over right."

"And you uh ..."

"Yeah, from the instruction register, is an or of a couple of things, will when I said bundle, there's actually two separate bunches of wires that come over from the instruction register, or they go into the controller."

"Our symbol for that is a slash ..."

"Oh."

"... and a number."

"Can you just draw that on there?"

"Well, we'd do that. I'd show you that — now how many lines?"

"There's actually two lines, both of which were 3 bits long, I think, two bunches of things."

"All right. Two of 3 bits, you say?"

"Right, oh, OK."

"Normally you just put it on a three; you wouldn't put bits."

"Right."

"... but identify which is, that you have 2 three-bit lines."

"OK, great."

"You just draw a slash on it and show you that it's 18 lines."

Ted: "Sure."

Claud: "... or whatever and just put down slash-18 and that tells you how many lines you got."

"Yeah, that's a nice notation."

"It's more definitive than bundle."

"Right."

"Well, I must admit, some of these things get simplified for drawing out ..."

"Yeah."

"... cause, you know like ..."

"It's easier to draw than; it's just a slash and the number on it. And I don't know where we picked it up, but we're not the only ones; It's sort of in the industry."

"Oh, OK, great."

"OK, controller, and all right. That, in essence, is modifiers to my control."

"Um-hum."

"... and now this does compare, minus greater than ..."

"That's actually shift to the right."

"Oh, shift right."

"Right."

"Shift right and add. All right. This one is compare, add and subtract."

"Mmm-hum."

"OK, now, can you do successive add/shifts here or multiply, although you've got a multiplier?"

"Yeah, I don't ..."

"Some designs can concatenate ..."

"You can actually concatenate through, um, if you're doing, uh, the stuff with the address ..."

"Mmmm-hum."

Ted: *"... you get to the point where you concatenate through two of these to go back out."*

Claud: *"Right."*

"But, I don't need that much in terms of add/shift because of my multiplier."

"Yeah, well, you multiply. By the time you get it in, you may have to add/shift all inside that box."

"Sure."

"But uh, yeah, all right. OK, by the way, normally if you concatenate adders, uh, you're clocking gets a little more complex, or you loose speed."

"OK."

"All right, because you don't need it very often, and it's easier to take the extra cycle, most of the time, then to mess up your clocking."

"Sure. Control is such a bug-a-boo that you really try to simplify that whenever you can."

"Yes, OK, now I see something. Oh, this is just a bucket of constants."

"Yeah."

"OK, yeah, you always need those, a lot."

"Do you have that normally in ROM or do you just normally burn that on the chip or do you use multiplexors to generate it, or how do you normally generate your constants?"

"It depends on the designer."

"OK, yeah."

"I have my preferences."

"What are your preferences?"

"Well with me, I go for the flexibility. I very seldom burn them in unless it's ..."

"OK."

Claud: *"Well, I have done more flexible designs than I have really honed for speed."*

Ted: *"OK."*

"You'll burn it."

"OK."

"OK, if you really are honing it for speed, like building vector processes, which I've never built, although I've consulted with Lowen, but with that you'll burn them in and all, because that is all — out speed — forget everything else, including cost almost. Now, these are a set of ..."

"Those are a set of single bit registers, uh, FEW what are they? Excuse me."

"Those things, they always focus at the wrong distance."

"Guaranteed."

"I can see close-up. It's reading signs down the road that I have a problem with."

"Those are error flags from the floating point information."

"Oh, OK, so these are my error flags. All right, I can ..."

"For floating point."

"... for floating point."

"Right."

"... and I presume ..."

"That would be my channel instruction register, but I didn't uh ..."

"OK, so you need those and hardware, no question, and, OK, channels."

"Those are my floating point registers."

"Yeah, OK, and this is a control register."

"Control registers."

"That's 16. This is just standard register, OK."

"MB is my primary memory."

"OK, all right. Now these represent ... You've got to get unaligned. Now this is a function for zero to 63."

Ted: "Right, especially there's some functionality about doing calculations to get some unaligned stuff and it will buffer that into a register. The reason why it buffers that into a register is because of the debugging ability. I want to be able to see if that functionality works properly. The real reason why there's a register there is because it goes over a clocked boundary I needed a place to put it."

Claud: *"OK."*

"... and that was the temporary register. I named it that so I would remember what it was."

"All right. In order to get the stuff on boundaries, why we have to do some shifting or cross-gating, or whatever."

"Yeah, because you don't know whether it's a boundary basis to start with ..."

"Yeah, it may be bits, and so you ... all right. This is for the same reason."

"Right."

"So, fundamentally or ..."

"To get and put for the PSW you do the same thing because the PSW is such a funny thing."

"Yeah, it's a lot of pieces, so you ... So those almost would be part of controls except that they're handling data ..."

"Right."

"... the object of some very complex control. And this is your first byte?"

"Actually that came about because the way the description was written, it was decided one would use a function in ISPS, which said return the first bit that's a one and how many bits over is it?"

"Oh, OK, oh, this is your first one, so you're going to count and tell me ..."

"Right."

"... the bit number of the first one."

"Right."

Claud: *"OK, fine."*

Ted: "That's the result of that."

"Got to have that. And uh ..."

"And that's the interrupt vector."

"Yeah, OK, and then we come to the instruction register, I presume."

"Right."

"And let's see, R94 three-two bitter."

"That has to do with the memory management stuff, as I recall. Let's see that has a segment table ..."

"OK."

"... and 32 bits a segment table."

"You got another one here."

"There's a page table."

"A page table."

"Right, and a segment index."

"Index."

"A page index, page displacement."

"Displacement."

"Right."

"And now we come to floating point."

"Now we start dealing with floating index stuff."

"Floating point stuff which uh, ..."

"Then there's the DAT box return itself."

"Mmmm-hum, DAT return, OK."

"All right, in terms of temporary registers, I end up having this temporary register and that temporary register between the two of them, in your structure, you would bring two busses into here."

"Yeah."

Ted: "What I did, basically is, I had one bus coming into here, and a bus coming into this which went into there."

Claud: *"Mmm-hum."*

"So like, if we follow this crazy line around ..."

"OK."

"We find that goes into the bottom of the ALU. This line here is that ..."

"Yeah."

"... let's trace that down. So that's ..."

"Oh, well, then maybe I ought to show you more on mine over here while I'm at it."

"All right."

"I had an AB register here."

"AH."

"Now, let me tell you, most of the time I would hold that gate down and read straight through it."

"OK."

"But if I'm clocking, I need it to hold at that point."

"Right, sure."

"But mainly — let me tell you what I use it for — was when I would clock here is so I could drop these lines ..."

"Right."

"... and use them for something else."

"Right, right."

"And this would still hold that through the adder so I could be latching."

"Right."

"That allowed me to latch, to read from, OK, my clock looked like this, that's not true, it's a balanced clock, but nevertheless I could read from, I could read on that edge, and I could write back in ..."

"Mmm-hum."

Claud: "... *later, onto that same clock* ..."

Ted: "Mmm-hum."

"... *provided I had a buffer.*"

"Right."

"*So, I made it a rule never to store in the capacitance. So that's the same thing.*"

"I don't store in capacitance either."

"*I don't believe in that.*"

"I think it's a crock."

"*Well, that's where you get intermittence.*"

"That's right. You get intermittence. Also, you're designing the chips manufacture, so you're in variability there."

"*Yeah, so, when I find anybody doing that, I write them off as a designer.*"

"That's right. That's what the purpose of that was."

"*OK, well, when I saw that, I realized I didn't show you those, two ...*"

"OK."

"*And that's what they're for is to hold it. And we have some other hold registers, I don't remember exactly where; my MAR, Memory Address Register, is on this 24-bit line. In reality, if you read into that, and come around, but functionally it's the same. I won't go into why we do that; it was for checking reasons, I can tell you. Yeah, by the way, over here, this actually goes up to this, and then we read out and back down. But that's so I can use the parity checker.*"

"OK."

"*Functionally, it did nothing for me.*"

"OK."

"*All right, I guess we've about almost finished, haven't we now?*"

"Yeah."

"*We've done that and this, they're buffers, and here's our prefix.*"

Section D.7   Back To Critique   195

Ted: "The prefix, the CPU timer, and the clock and the time of day clock."

Claud: *"OK, that's just the architectural stuff that you've gotta have."*

"You had those out in the stack also?"

"Yeah, I showed them out; I just didn't have room ... And there's the signal out here."

"Signal out, I/O done."

"There's a one-byte job'ie."

"Yeah. It's R-67 here, you're indexed that way."

"Yea, I am indeed am; this is high-class operation. Written out by hand is not as high-class as I would like it. But R-67 is a signal output for direct I/O. That's, uh ..."

"Oh, OK."

"... ten bits out. So you had eight bit out ..."

"Yeah."

"... and I had ten bit out."

"Yeah, all right."

"I had eight bit, the next I/O data register. That's eight bits ..."

"Yeah."

"... and then the next up is external-register; bit zero is the timer interrupt; bit one is the console interrupt, uh ..."

"Mmm-hum."

"... next up is device register; that holds the device address as zero to 255."

"OK."

"Then the channel address, channel instruction and channel condition codes ..."

"Yeah, OK, that's all in my channel stuff."

"Right, that you had out in the architectural register?"

"Mine was out, uh ..."

Ted: "On a separate chip."

Claud: *"This chip again in a different implementation."*

"Actually, I have the design for that stuff, but I decided not to draw it because I didn't see it anywhere in your stuff."

"No, it's not. No, I didn't show it, no. What's this?" Oh, that's a channel interrupt?"

"Interrupt, or instruction, I don't know. It must be interrupt."

"Instruction, I'm sorry. Is that 62?"

"It's 32."

"Yeah, it's channel interrupt."

"Channel interrupt."

"Request."

"And there's a PSW and PSEC."

"It's 31. That's the extended code stuff."

"I've got ya, yeah."

"Where'd you do your extended code stuff?"

"Eh, yeah we handled that out of board. We had that, there was actually two registers. Yeah, we carried two registers, that was in the stack."

"Right, OK."

"Memory buffer registers, OK."

"Buffer, double-buffer, and address register."

"Yeah, um-hum, OK, I had those um, on memory address register."

"By the way, there was a ROS address down here, too, if you want all that."

"Sure."

"We have, there's a ROS data register, ROS data register there, that was on chip, and there's a ROS address register that came down here as well. And, let's see, there was a comparator ..."

"The ROS address was also coupled to your ..."

Claud: *"Oh, it had all these bits and so on, and the ALU; there's a whole complex mess, uh, because that is a critical path for speed."*

Ted: "Yeah."

"And so you'll find that we bypass some of that to get free-charging on lines and all that sort of stuff. But that's uh, technology dependent."

"Sure."

"I don't consider that as being ... that's the tough engineering part. That has nothing to do ..."

"Yeah."

"... with architecture."

"Now, we also did our DAT compare right here; I don't remember whether I did it in this, I did, yeah. But I had a DAT compare function here, only to compare. The rest of it, I just didn't have space."

"I see. You put your compare right there, OK."

"Yeah. Uh, and then the rest of it was done in microcode, simply because we ran out of blocks."

"This is, could you do me a favor and plug it in that plug over there? In talking between the two, I don't want to lose half the conversations."

"Yeah, my voice carries, and normally ..."

"Mine carries; I think this is a pretty good tape recorder in that the microphone was a good design."

"Um-hum."

"And uh, I've had good luck with it so far."

"It looks like a nice one. Uh, I don't know much else about that. I think I've pretty well covered the little bit I had."

D.8 Summary

"Well, let me try summarizing."

"OK."

"... some of the key points. Uhm, one is that in general, the registers I have here, you indeed have them mostly out board on the stack."

Claud: *"Right."*

Ted: "Uhm, you have primarily an 8-bit data path with several registers to go through for reasons of cost and there wasn't an increase in performance. Uh, you have the single ALU; I have multiple ALUs. Both, again because it's a Cadillac version, more higher performance."

"I do have a multiple; one for my address."

"Yeah OK."

"I have an address, which is a 24 bitter, and then an 8 bit."

"And I have three, so, I have one more."

"Yeah."

"You fold your uh, logical operators into your ALU."

"Right."

"... and you mentioned that they possibly would have it out for debugging, but you like it in for cost reasons. OK?"

"Correct."

"Uh, let's see, uh, constants are the same; so pretty much of the design is pretty much the same sort of design as yours, with exception of the bit width, uh, being wider on mine than on yours."

"Right."

"OK."

"You may rearrange some gating between the registers at the time you start laying out your busses."

"Sure."

"But that is technology dependent."

"Sure."

"... and so you don't make that decision."

"When you said you had a Christmas tree four-part bus, what did the bus actually look like?"

"That was for driving capability and it looked like this."

Ted: "OK."

Claud: *"OK, I had an OR at this point — an OR circuit — and that was really the bus, OK."*

"Uhm-hum."

"This is your bus; we're sure of how to get there. And then I would take this many, well, let me see. I guess I had a driver, and it drove into that. I had four of these."

"Uh-huh."

"Uh, excuse me, not so — that's the driver down below. Let me get this straightened out; I would take part of my registers — you can only do — and I'd put 'em in this way. My next set of registers, they're here; would go into the next one, and uh, so on, and then I'd have four of these lined up to take care of my 17 registers."

"Sure."

"OK, actually, I don't want to put 16 in, so I divided it into four fours is what it boiled down to. That's one, two, three, four; one, two, three, four; one, two, three, four."

"OK."

"And so I would handle 16 registers, and these then, would go through the OR, and it was all for powering reasons like that."

"Right."

"And then this turned out to be the bus. The others would be, uh, where I was using only two of those, and then to an OR ..."

"Um-hum."

"... and this would be for your registers. And that's the way we did that."

"OK."

"Uhm, let's see."

"That's what I meant by Christmas tree."

"Sure."

"And that was for powering and to break that capacitance up into pieces."

Ted: "What is your overall impression? If I were a designer coming to you with this as a first-shot at a design, and saying that my constraints were for a high performance machine, uh, would you think this is an OK, machine?"

Claud: *"It's a good starting point because ..."*

"... you can't go much further without knowing your technology."

"Sure."

"Technology, your package, and decide uh, uh, then, because, you see, take this Christmas tree thing over here."

"Yeah."

"That doesn't show up logic-wise at all ..."

"Sure."

"... anywhere. That's just a straight line."

"That's right."

"And then you've got to add all that in, because of the technology I was working with."

"Right."

"And so, that'll work."

D.9 Parting

"Now, see, in terms of ... you said you might be able to send me some stuff. I'd be happy to give you my address or whatever. You know, if you're going to throw out any notes ..."

"Yeah."

"... or, you know, uh ..."

"I've got one right on top of the thing there, that's mine. Yeah I can give you that. And, they ought to ... if you don't have access to a data flow ..."

"No, I ... yeah."

"... for one of our machines."

"Yeah, I'd like one of those, too."

Claud: *"I'll send you one just to ..."*

Ted: "That's good."

"Because they are more optimized as to what you'd actually ... well, they're actually what you see in industry out ... we sell that."

"Um-hum."

"This was an experiment. Who's our representative to Carnegie-Mellon? Is that uh, Dasgupta?"

"He's the one who's got, uh ..."

"Well, if so, I should go through him rather than you dealing with a whole bunch of people."

"Right."

"Uh, we find out that, uh, that doesn't seem to work, if you should get a bunch of us working directly than not talking to each other."

"OK, now that's my address. However, I would recommend not sending it to me, but sending it to my advisor, and I'll give you his address. The reason for it is, my advisor ... see, professor's mail gets shelved in one box, in a box individual for professors. Graduate students' mail is thrown in a heap."

"Oh, I see."

"And, if some other graduate student saw something from IBM that he thought was interesting, he might just pick up the stuff and walk away with it. I may never see it. And also, I'll give you my ... Don Thomas' phone number because he has a secretary who picks up and answers the phone, whereas I have ... I guess I have eight office mates, and maybe they'll give me the note, and maybe not."

"Yeah."

"So you know, life for a graduate student isn't as reliable as life for a professor. But, at any rate ..."

"That's right."

"... so, most correspond ... I'd prefer if you would send it to Don instead, but this is mine anyway."

"All right, well, I'll send it through you."

Ted: "OK."

Claud: *"We only want one interface anyway."*

Sumit: "I average about a call a month anyway."

Ted: "Yeah. I, us ..."

Sumit: "I apologize for running around. Uh, I had several things going at the same time."

"Do you want me to really hurt him?"

Ted: "Go ahead, hurt him."

"We didn't miss you. How's that?"

Sumit: "I hope not. I am not very knowledgeable about the subject. The person who is knowledgeable is sitting right here."

Ted: "Yeah."

Unknown: "Not on that."

Sumit: "She's knowledgeable on the blue book."

Ted: "Yeah, um-hum."

Unknown: "Where'd you get that, I meant to ask?"

Ted: "Oh, I'd be happy to give you as many copies of it as you'd like."

Unknown: "No, no, no."

Ted: "No, where'd I get it? Um, there actually were two people who, at Carnegie-Mellon, who spent their Masters Degree doing it."

Unknown: "Yeah, I saw that."

Ted: "Uhm ..."

Unknown: "Where originally, did that come from?"

Ted: "Oh, this actually, originally was, they sat down with the Bible ..."

"... with the Prince Ops."

"... and, uh the principle of operations and code it one-by-one."

Unknown: "That's what I was looking for."

Ted: "And they code it ..."

Section D.9   *Parting*

Unknown: "You just can't, I mean, anybody off the street's just not going to know it."

Ted: "No, if you're not familiar with Prince Ops., that doesn't mean anything."

Unknown: "And what they did with it ..."

Claud: *"Because it's a translation of a very precise definition."*

Unknown: "I see. OK."

Ted: "And they also did it at the channel controller and uh ..."

"OK, we already have all these addresses and stuff don't we?"

Sumit: "Yeah, we do."

"Well, I'll just look and see, but anyway I'll keep it, but, uh, I'll look and see if I can't, uh, give 'em a little more detail on mine, on what we did from here."

Ted: "What we did, data flow."

Sumit: "But, uh, now I don't know whether I can lay hands on that. You can, I suppose around here."

"Yeah, around here."

Sumit: "But, they're in the public domain."

"Yeah."

Sumit: "Necessarily, it's not a ..."

"OK."

Ted: "So, the two action points are more info on the ICCC paper stuff ..."

"Yeah."

"... and uh, the ..."

"Yeah, I'm going to do that and the 158 would be a good one for him to look at. It's uh, been around for years anyway." "Excuse me. OK, can I make a copy of that document?"

"I may insist."

"Uhm."

Unknown: "It's on here."

Ted: "... would you like to, uh, well, I guess that's about it. I thank you very much for your time, Claude."

Claud: *"Well ..."*

"It's been a pleasure meeting you and ..."

"It's a pleasure."

"... and ..."

"I hope it'll be a little helpful."

"Oh, it's been very helpful. You see, what I've done is I worked a lot with two designers on 6502."

"Yeah."

"And I really optimized my rules for the 6502."

"Yeah."

"And so, the question that's asked is uhm, well, have you learned about doing design in general or have you learn about designing 6502? And, so ..."

"Fair question."

"... which is a fair question and it's a good point. So, uh, the reason why I put this through my system is to say, well look, not only if I scaled up I've gone up by over order of magnitude, OK, but I've also done a design I've never seen before. And now I've worked with a designer who I've never seen before."

"Yeah."

"And he thinks it's a reasonable design. Dot, dot, I have really learned something about ... my system has really learned something about doing architectural designs in general. And, uh, that's a real important data point Claude. I thank you very much for it."

"Mmmm-hum."

"Uhm, that you will probably be more than a whole chapter in my thesis."

"But uh, well it's, you know ..."

Sumit: "Good."

Ted: "... it's a good piece of ..."

Claud: *"No, I have found that uh, if you can handle a 370 in the hardware side, that uh, the less complex seem to work pretty well."*

"Uhm-hum."

"You wind up with uh, things you don't use ..."

"Sure."

"... but uh,"

"Sure."

"... you can still do 'em."

"One of the interesting things that the 370 brought up was ... the 6502 has one bus, OK."

"Yeah."

"And one bus, which is eight bits for everything except for the program counter which is uh, 16 bits."

"OK."

"So, in a program counter you just have a couple of discreet wires rather than it being actually a bus. Well, this actually broke down to an 8-bit bus, a 64 bit-bus and a 24-bit bus."

"Mmm-hum."

"And it's interesting to note that you had a 24-bit bus, and you had an 8-bit bus, and I had the 64 bytes, and the only reason why I had the 64-bit bus is because I didn't serialize those operations into, or you know, load, load, load."

"Yeah, OK."

"My system actually handled that, which was nice. It never had to handle that before."

"Now, if I were going for performance, I would have had a 64."

"Right."

"And uh, a lot of our machines do."

Ted: "Sure."

Claud: *"And, uh, some of 'em are even bigger than that."*

Sumit: "And in fact, this design is aimed at uh, performance."

"Right."

Ted: "Yeah, uh, these are ..."

Sumit: "From that standpoint ..."

"He is designing a step above."

Sumit: "So from that standpoint, there's agreement then, between these two pretty well?"

"Oh, yeah."

Sumit: "That's good."

"Yeah, the data transmissions are very important."

Ted: "Well, I'm very happy, very happy. Uhm, I can sit and chat with you more about expert systems if you want, if you're ... I don't know about what's ... If you have questions or, I can show the real rules; if you want to see what the real rule looked like, and that English version is a bit doctored uhm ..."

"I may be going to dash away then."

Sumit: "Thank you very much."

"Thank you again, Claude."

Sumit: "It was nice of you to come by. By the way, one question."

"Yeah."

Sumit: "You probably have a better knowledge than I of the people you go to for the data flow."

"All right."

Sumit: "People like uh ..."

"Uh, there's a product engineering group that's doing that, and I don't know their names right off, uh."

Sumit: "Mmm-hum."

Claud: *"Uh, I don't know. But uh, we have a group maintained, you know, a group of about five or six people, and uh, let me give you a call."*

Sumit: "OK."

"I'll find out who it is."

Sumit: "All right. Fine."

"In terms of this ..."

Sumit: "Ask them questions, too."

Ted: "This is OPS5, the language I program all my stuff in is OPS5 and uh, what I really ... thanks again, Claude."

"Yeah. Good luck."

"Thank you." "Thank you Claude, once again."

INDEX

abstract representation 3
acquisition
   interview 16-18, 26-27, 39-40
   knowledge 12, 17, 19, 26-27, 40, 104
   method 17-19
   system, knowledge 105
action
   modification 32
   rule, situation 16
active
   context 25, 32-39, 55, 63, 65, 70, 74, 85
   goal 38, 55, 85
   rule 34-38, 85
adder, module 25
addition knowledge, incremental 26, 39, 87, 103
address
   bus 93-96, 98
   carrier, page 44
   local store 98
   memory 65, 72, 94, 96, 99, 102
   port 24, 65, 72, 94-95, 102
   register 24, 72, 93-94, 96, 98-99
   virtual 98, 101
algorithm, synthesis 7
algorithmic
   description 1-4, 6, 24, 28, 41-42, 63, 101-102
   operator 2, 6, 28, 41
allocate
   clock 22, 40, 66
   hardware 6-7, 23-24, 35-36, 39, 47, 63, 66, 68-71, 76, 90, 104

memory 65, 69
module 24, 35, 55, 58, 64-68, 70-71, 76, 80, 82-83
multiplexer 7, 55, 83, 85
register 7, 22, 24, 33, 40, 58, 63-68, 70, 73, 75, 80, 82-83
allocation
   bus 32, 83, 85-86
   control 4, 18, 22, 32, 34-35, 40, 66, 68, 70, 86, 104
   design 4, 6, 18, 22, 39-40, 59, 66, 73, 82, 86, 103-104
   fold 33-35, 74
   global 18, 39, 58-59, 61, 63-64, 66-67, 72, 82, 86, 104
   goal bus 85
   goal fold 73
   goal operator 70-71
   implementation 58, 104
   local 86, 104
   rule 36, 59, 61, 68, 70-71, 82
   subtask 58-59, 61, 103
   task 17-18, 39, 61, 66, 86, 103-104
   temporary register 70
   variable 34-36, 65, 70
   VT 58, 63, 67, 72, 82
ALU 33, 48, 58, 71, 94, 99
antecedent
   reasoning 11, 26
   rule 11, 14, 26, 32, 38-39, 87, 105
architectural
   module 24, 70, 86, 104
   register 24, 52, 70, 86, 90, 93-94, 98, 102, 104
architecture, System/370 93, 98-99, 101

INDEX

area, design 2-3, 7, 22, 40, 48, 82
arithmetic
  operation 42, 93, 96
  type 35
array
  masterslice gate 96
  memory 90, 94, 99, 102
  register 65, 79, 90, 93-94, 99, 102
artificial intelligence 2, 9, 15
assignment
  control 4, 44, 48, 55, 63, 67-69
  operator 4, 24, 44, 48, 55, 58, 67-69
attribute
  module 25, 62, 66-67, 70-72, 75-76, 85
  register 33, 67, 71-72, 75
  type 25, 44, 66, 75-76
  value pair 10, 25
average, weight 78-79, 81-82
backtracking 2, 5, 7, 23, 26, 40, 102-103
backward chaining 11
base
  knowledge 5, 16-17, 24, 26, 105
  rule 33, 105
based expert system, knowledge 2, 4, 7, 9-10, 12, 15-16, 23, 31, 40, 103-105
behavior, expert 18
Bell Laboratories 27, 71, 77
bind
  module 4, 24, 33, 73, 75-76, 85, 102
  register 4, 7, 24, 73, 75
block
  functional 90, 96, 98-99, 101

select 39, 45, 66
book knowledge 16-18, 23, 40
bookkeeping 41, 55, 61-62, 64, 84
branch
  input 44, 53, 71
  operator 44, 52, 71, 76, 90
buffer register 94, 96, 98, 101
bus
  address 93-96, 98
  allocation 32, 83, 85-86
  allocation, goal 85
  communication 22
  configuration 85-86
  example 3, 44, 83, 85, 101
  fan-out 58, 83, 86, 96, 98
  multiple 32, 58, 86
  port 55, 84-86, 94-96
  rule 29, 33, 83, 85
  structure 22, 83
  style 3, 22, 90, 93, 101
  usage 85
cache scheme 101
calculation, stable 38
call, procedure 32-33, 38-39, 44
capability, hardware 79, 81
Carnegie-Mellon University 1, 89
carrier
  instruction 44
  page address 44
  temporary 42, 44
case study
  interview 16, 18, 20, 39
  method 18-19
chaining
  backward 11
  forward 11, 26
change
  context 35, 69
  rule 27, 29, 35, 37, 87, 102, 105

technology 22, 40, 87
chip
    silicon 32
    working 32, 99-100
choice, design 23, 40, 90, 101
clean, goal 69, 76, 84
cleanup
    fan-out 58, 85-86
    goal 73, 86
    multiplexer 58, 82-86
    rule 32, 58, 73, 76
    working memory 32, 73, 76
clock
    allocate 22, 40, 66
    cycle 93, 96, 102
    phase 18, 22, 40, 66, 93
    step 69
    synchronization 50
    two phase 93
CMU/DA 3
code
    register 94, 98
    status 34, 94
codified knowledge 5, 16, 23, 31, 40, 63
combination, link 71
combine operator 28
communication bus 22
compiler optimization 24, 44
complexity, design problem 69
component
    discrete 90, 93
    minimize 87, 104
    select 14-15
compound operator 44-45
concatenation module 67, 75, 85
configuration
    bus 85-86
    multiplexer 85-86

connection
    minimize 7, 21, 39, 87
    route 41
connectivity 3, 27, 75-76, 87, 104
consequent reasoning 11
consideration, technology 7, 21, 39, 47
constant
    module 24, 33, 64, 66-67, 71, 84
    technology-sensitive 87
    unused 83
constraint
    default 35, 64, 66
    design 2, 6-7, 21-23, 26, 35, 39-41, 47-48, 56, 63-64, 69, 76, 87, 100-102, 105
    handling 104
    information 35, 47-48, 64
    problem 2, 6-7, 23, 96
    supplied 2, 64
    technology 7, 18, 21-22, 39-40, 47-48, 64, 100
    violation 18, 22, 24, 40
contents, module 62, 77, 80, 82
context
    active 25, 32-39, 55, 63, 65, 70, 74, 85
    change 35, 69
    goal 38, 55, 61
    pending 34, 55
    status 34, 55
    switch 72
control
    allocation 4, 18, 22, 32, 34-35, 40, 66, 68, 70, 86, 104
    assignment 4, 44, 48, 55, 63, 67-69
    management 58, 61

INDEX 211

register 2-5, 18, 22, 24, 38, 40, 66-68, 70-72, 74, 78, 86, 94, 96, 104
sequence 2, 34, 52
specification 52-53, 63, 67, 95
step 24, 32, 34-35, 38, 44, 47-48, 52, 55, 63, 67, 69, 72, 74, 76, 86, 104
step, micro 48, 55
structure 22, 32, 40-41, 52, 69
task 34, 41, 61, 66, 86, 103-104
controller
  design 21, 52, 63, 66, 90
  memory 90
  module 52, 55, 64, 66-67
  operator 52, 71-72
  programmable 21
cost
  estimator 27, 41, 73, 75-77, 80-81
  module 33, 48, 67, 71, 75-76, 80-82
  operator 6, 27, 33, 73, 78-79, 81
  port 81-82
counter, program 44, 98
create module rule 61
creation, database 4, 35
criteria, selection 31
cycle
  clock 93, 96, 102
  match 33
D370 design 90, 99-101, 103
data
  dependency 6, 69
  flow 3, 6, 24, 30, 96, 101
  operator 2-6, 18, 22, 25, 27, 40, 42, 66, 69-70, 72, 77, 86, 90, 104

path 2-7, 18, 22, 27-28, 40, 59, 62, 66, 70-72, 77, 80, 86, 90, 93, 99, 101, 104
register 2-7, 18, 22, 28-30, 40, 66, 72, 86, 98-99, 101, 104
representation 6-7, 24, 31, 57
database
  creation 4, 35
  operator 4, 33, 35, 69
  technology 47-48, 58, 64, 66
  technology-independent 4, 32, 35
  technology-sensitive 66, 69
Davis, Claud 90, 99, 102
decimal
  instruction, packed 89
  number, packed 96, 99, 101
declaration, map 90
declared register 29-30, 63, 65, 90
decode loop 44, 47, 52, 64, 67, 73, 82
default constraint 35, 64, 66
delay
  information 24
  step 35, 47-48, 69
  value 35, 47, 50, 69
demon rule 37
dependency, data 6, 69
description
  algorithmic 1-4, 6, 24, 28, 41-42, 63, 101-102
  implementation 4, 42, 52, 82, 102
  input 24, 32, 37, 42, 90
  language 3, 41-42, 52, 56
  language, hardware 42
  network 24
  system 1-2, 42, 52
  technology-independent 2, 24, 32, 41, 52

VT 89
design
    allocation 4, 6, 18, 22, 39-40, 59, 66, 73, 82, 86, 103-104
    area 2-3, 7, 22, 40, 48, 82
    choice 23, 40, 90, 101
    constraint 2, 6-7, 21-23, 26, 35, 39-41, 47-48, 56, 63-64, 69, 76, 87, 100-102, 105
    controller 21, 52, 63, 66, 90
    D370 90, 99-101, 103
    environment 7, 102, 104
    feasibility 21, 39
    hardware 3-4, 6, 21, 23, 36, 39, 50, 66, 69, 74, 84, 87, 104
    implementation 2, 5, 7, 18, 32, 102, 104-105
    iteration 22-23, 40, 90
    knowledge 2, 4-5, 7, 15-16, 18, 26-27, 31-32, 39-40, 61, 63, 89-90, 102-104
    method 2-3, 17-18, 102
    microprocessor 3, 21, 31, 40, 50, 89
    non-pipeline 69
    parallel 6, 18, 22, 24, 39-40, 48, 69
    problem complexity 69
    process 1, 3, 16, 18, 22-23, 40, 84
    processor 3, 5, 50, 89, 101, 104
    register 6, 21-22, 24, 27, 29, 33, 40, 52, 63, 66-67, 72-75, 79, 82, 84, 86, 90, 99, 103-104
    rule 4, 26-27, 29, 31-38, 40, 58-59, 61-63, 66, 69, 82, 84, 90, 101, 105
    rule shrink 23, 40
    serial 18, 22, 39-40, 76
    space 15, 18, 26
    style 3-4, 6, 18, 23, 40, 90, 104
    synchronous 74, 87, 104
    synthesis 1-7, 18, 24, 44, 75-76, 95, 102-104
    system 1-7, 15, 18, 21, 23, 26-27, 87, 89, 102-105
    System/370 5, 7, 90, 100, 104
    team 21, 90, 99
    technique 4-5, 18, 21, 26, 35, 69, 102-103
    technology 1, 3, 7, 21-22, 39-40, 48, 99-100
    time 1-2, 18, 27, 35, 88, 90, 102-104
    tool 1, 3-4, 7, 15, 23, 61, 103
    usable 4, 15, 104
designer performance 77, 80, 82
destination port 25, 62-63
development
    summary 39
    system 1-3, 7, 9, 15-16, 39-40, 105
diagnostic system 12
difference, set 81
digital hardware 77
discrete
    component 90, 93
    multiplexer 85
discrimination net 105
division, problem 41, 58
documentation, multi level 1, 104
domain
    knowledge 4, 9-10, 12, 14-15, 31-32
    problem 5, 9-10, 12
    system 4, 9-10, 12

INDEX

domain-independent
  knowledge 32, 37
  rule 36-39
domain-specific knowledge 5, 26, 32
element
  logic 93
  match 11, 26, 47
  storage 6, 24, 63
elicitation procedure 17, 27
engine, inference 31
environment
  design 7, 102, 104
  synthesis 7, 104
  teaching 105
error
  parity 98
  situation 16
estimator 27, 41, 73, 75-77, 80-81, 87
  cost 27, 41, 73, 75-77, 80-81
  layout 38
  partition 27, 41, 73, 75-77, 80, 87
  proximity 75-76
example
  bus 3, 44, 83, 85, 101
  pipeline 3
  problem 9, 16
execution time 87, 103
expand, memory 31, 40, 103
experiment
  partition 77, 80
  price 82
  System/370 31, 102
expert
  behavior 18
  knowledge 2, 4-5, 9-12, 14-17, 23, 26, 40, 58, 87-88, 103-104

system, knowledge based 2, 4, 7, 9-10, 12, 15-16, 23, 31, 40, 103-105
explicit knowledge 4-5
fabrication technology 1
Facet 6
factor
  technology-sensitive 77, 79-80
  weight 77, 79-80
fan-out
  bus 58, 83, 86, 96, 98
  cleanup 58, 85-86
feasibility, design 21, 39
feeding, register 33, 74, 84, 93, 96
field, microcode 96
firing, rule 31, 64, 82
floating point 89, 94, 99
floor plan 15, 58, 77, 80, 82, 87, 101, 104
flow, data 3, 6, 24, 30, 96, 101
fold
  allocation 33-35, 74
  allocation, goal 73
  parameter 35
formal
  input 44, 90
  parameter 44, 63, 66, 84
forward chaining 11, 26
function
  logic, single 93
  management 63
  service 58-59, 61, 64, 71
functional
  block 90, 96, 98-99, 101
  knowledge 31
  pair 98
  subtask 58
gate array, masterslice 96
general purpose register 102

generator, parity 96, 98
global
    allocation 18, 39, 58-59, 61, 63-64, 66-67, 72, 82, 86, 104
    improvement 7
    storage 63
goal
    active 38, 55, 85
    bus allocation 85
    clean 69, 76, 84
    cleanup 73, 86
    context 38, 55, 61
    fold allocation 73
    make link 63, 75
    make module 61
    module mv 61, 75-76
    module rm 62, 75
    operator allocation 70-71
    select 61
    unreferenced 84-86
group, register 96, 98, 102
handler, trap 98
handling, constraint 104
hardware
    allocate 6-7, 23-24, 35-36, 39, 47, 63, 66, 68-71, 76, 90, 104
    capability 79, 81
    description language 42
    design 3-4, 6, 21, 23, 36, 39, 50, 66, 69, 74, 84, 87, 104
    digital 77
    implementation 4, 78, 80, 100
    module 24, 28, 35, 68, 71, 75-78, 80-81, 102, 104
    multiplexer 3, 24, 28
    non-changing 63, 66
    opcode 35, 47-48
    operator 6, 28, 35, 47, 66, 70, 77-79, 90, 104
    price 80, 87
    register 6-7, 24, 66, 68, 74-75, 96, 104
    share 28, 77-79, 81, 87
    technology-specific 102
    translation 101
    usage 6, 90
hardware-network level, technology-independent 5, 77, 80
high level
    overview 21, 39
    partition 39, 41, 58, 80, 87, 90, 101
implementation
    allocation 58, 104
    description 4, 42, 52, 82, 102
    design 2, 5, 7, 18, 32, 102, 104-105
    hardware 4, 78, 80, 100
    parallel 44
    simulator 7
    style 77-78, 80
    synthesis 2, 7, 18
improvement, global 7
incremental addition knowledge 26, 39, 87, 103
inference engine 31
information
    constraint 35, 47-48, 64
    delay 24
    partition 24, 27, 41, 58, 77, 80
    technology-sensitive 47-48, 77, 80
    time 2, 20, 86
initial
    knowledge 17, 23

INDEX

state 26, 83
input
   branch 44, 53, 71
   description 24, 32, 37, 42, 90
   formal 44, 90
   multiplexer 24, 28, 52-53, 55,
      62, 67, 82, 84-85
   operator 24, 33, 37, 44, 52,
      70-73, 75, 90
   port 24, 52, 55, 62-63, 65, 67,
      70-72, 74, 82-85, 94, 96,
      102
   register 24, 28, 33, 48, 52, 65,
      67, 70, 72, 74-75, 82, 84,
      93-94, 96, 98
   specification 37, 52-53, 63
   trim 70, 72
   value 33, 44, 53, 70, 74, 84
instruction
   carrier 44
   logic 93
   packed decimal 89
   register 45, 67, 93-94, 96, 99
   unpack 96
intelligence, artificial 2, 9, 15
interaction, user 104
interface port, module 52
internal representation 24, 87
interpreter
   pattern 11, 25-26
   rule 10-12, 14-15, 25-26
interview
   acquisition 16-18, 26-27, 39-40
   case study 16, 18, 20, 39
   method 17-18, 58
   technique 5, 18, 26
ISPS 3, 42
iteration, design 22-23, 40, 90
Jaccord 78

key, storage 94
knowledge
   acquisition 12, 17, 19, 26-27,
      40, 104
   acquisition system 105
   base 5, 16-17, 24, 26, 105
   based expert system 2, 4, 7, 9-
      10, 12, 15-16, 23, 31, 40,
      103-105
   book 16-18, 23, 40
   codified 5, 16, 23, 31, 40, 63
   design 2, 4-5, 7, 15-16, 18,
      26-27, 31-32, 39-40, 61, 63,
      89-90, 102-104
   domain 4, 9-10, 12, 14-15,
      31-32
   domain-independent 32, 37
   domain-specific 5, 26, 32
   expert 2, 4-5, 9-12, 14-17, 23,
      26, 40, 58, 87-88, 103-104
   explicit 4-5
   functional 31
   incremental addition 26, 39,
      87, 103
   initial 17, 23
   link 62-63, 71
   modular 39, 103
   representation 7, 24, 31, 58
   task 5, 7, 11, 15, 17, 26, 32,
      39, 103
   type 5, 15, 31-32, 58, 61
   world 31, 40
language
   description 3, 41-42, 52, 56
   hardware description 42
   processor 3
   programming 3, 6-7, 10, 15,
      25, 42, 77
   specification 7, 52

layout
    estimator 38
    program 69
learning process 40
level
    documentation, multi 1, 104
    overview, high 21, 39
    partition, high 39, 41, 58, 80, 87, 90, 101
    synthesis 3-4
    technology-independent hardware-network 5, 77, 80
link
    combination 71
    goal, make 63, 75
    knowledge 62-63, 71
    make 55, 62-63
    management 58-59, 62-63
    output 24, 52, 55, 62-63, 67, 70-72, 74-75, 84-86
    request 55, 62
    rule 33, 55, 59, 61-63, 70-71, 84-85
list
    operation 32, 38
    simulate 32, 38
    working memory 32, 38, 46-47, 50, 54, 61, 69
local
    allocation 86, 104
    store 41, 99, 102
    store address 98
logic
    element 93
    instruction 93
    single function 93
loop, decode 44, 47, 52, 64, 67, 73, 82
make

link 55, 62-63
link goal 63, 75
module, goal 61
management
    control 58, 61
    function 63
    link 58-59, 62-63
    module 58, 61
    operation, memory 89
    port 58-59, 62-63
    rule 59, 61-63
    service 63
    storage 98
map
    declaration 90
    step 44
masterslice gate array 96
match 4, 11, 26, 33-34, 47, 76-77, 80, 82, 102-103
    cycle 33
    element 11, 26, 47
    rule 11, 26, 33-34, 47
    state 26, 34
    technique 102-103
measure, similarity 78
mechanism, reasoning 12, 26, 87
membership, set 77, 87
memory 10-12, 14-15, 21, 25-28, 31, 35-37, 40, 61, 65, 69, 72, 76, 87, 90, 94, 96, 98-99, 101-103
    address 65, 72, 94, 96, 99, 102
    allocate 65, 69
    array 90, 94, 99, 102
    cleanup, working 32, 73, 76
    controller 90
    expand 31, 40, 103
    list, working 32, 38, 46-47, 50, 54, 61, 69

INDEX 217

management operation 89
module 61, 65
primary 94
remove working 32, 36, 69, 73
representation, working 24,
 46-47, 50, 54-55, 87
rule 10-12, 14-15, 25-26, 31,
 35-37, 40, 61, 69, 87, 90,
 102-103
structure, working 55, 69, 87
system 10-12, 21, 87
working 10-12, 15, 21, 25-26,
 31, 35-37, 61, 69, 76, 87, 96
method
 acquisition 17-19
 case study 18-19
 design 2-3, 17-18, 102
 interview 17-18, 58
 weak 17, 26
micro control step 48, 55
microcode
 field 96
 special 96, 98
 symbolic 24, 95
 word 95, 98
microcontroller 93-94, 98
microprocessor, design 3, 21, 31,
 40, 50, 89
minimize
 component 87, 104
 connection 7, 21, 39, 87
model, specification 52
modification
 action 32
 rule 33, 37, 90
 time 33
modular knowledge 39, 103
module
 adder 25

allocate 24, 35, 55, 58, 64-68,
 70-71, 76, 80, 82-83
architectural 24, 70, 86, 104
attribute 25, 62, 66-67, 70-72,
 75-76, 85
bind 4, 24, 33, 73, 75-76, 85,
 102
concatenation 67, 75, 85
constant 24, 33, 64, 66-67, 71,
 84
contents 62, 77, 80, 82
controller 52, 55, 64, 66-67
cost 33, 48, 67, 71, 75-76,
 80-82
goal make 61
hardware 24, 28, 35, 68, 71,
 75-78, 80-81, 102, 104
interface port 52
management 58, 61
memory 61, 65
mv, goal 61, 75-76
non-bus 85
physical 77
register 4, 24, 48, 52, 58, 64-
 68, 70-72, 75, 80, 82-84, 86,
 104
remove 61-62, 66, 72, 75-76,
 83-84
rm goal 62, 75
rule, create 61
rule, move 61
technology-independent 4, 24,
 80
temporary 67, 70, 75-76
temporary register 70, 75
type 25, 28, 33, 52, 61, 66, 76,
 80, 84
unbind 33
unreferenced 83-85

unused 82
wiring 52, 67, 82
move module rule 61
multi level documentation 1, 104
multiple
  bus 32, 58, 86
  rule 26, 29
multiplexer
  allocate 7, 55, 83, 85
  cleanup 58, 82-86
  configuration 85-86
  discrete 85
  hardware 3, 24, 28
  input 24, 28, 52-53, 55, 62, 67, 82, 84-85
  number 30, 55
  tree 83-84
multiplication operator 37, 55, 70
mv, goal module 61, 75-76
MYCIN 14
name translation 47
net, discrimination 105
network
  description 24
  rete 87
non-architectural register 24, 79
non-bus module 85
non-changing hardware 63, 66
non-pipeline design 69
number
  multiplexer 30, 55
  packed decimal 96, 99, 101
opcode, hardware 35, 47-48
operand register 98
operation
  arithmetic 42, 93, 96
  list 32, 38
  memory management 89
  select 53

set 32, 38, 44
shift 24, 42, 71, 98, 100
synthesis 24, 102, 104
transform 32, 71
wiring 71, 73, 84
operator
  algorithmic 2, 6, 28, 41
  allocation, goal 70-71
  assignment 4, 24, 44, 48, 55, 58, 67-69
  branch 44, 52, 71, 76, 90
  combine 28
  compound 44-45
  controller 52, 71-72
  cost 6, 27, 33, 73, 78-79, 81
  data 2-6, 18, 22, 25, 27, 40, 42, 66, 69-70, 72, 77, 86, 90, 104
  database 4, 33, 35, 69
  hardware 6, 28, 35, 47, 66, 70, 77-79, 90, 104
  input 24, 33, 37, 44, 52, 70-73, 75, 90
  multiplication 37, 55, 70
  outnode 44, 70, 76
  output 24, 33, 37, 44, 70, 72
  parallel 6, 24, 42, 69, 81, 90
  proximity 33, 78-80
  specified 27, 42, 44, 69
  trim 37, 55, 58, 69-70, 72
  type 25, 27-28, 37, 44, 70, 77, 79
OPS5 5, 12, 23
optimization, compiler 24, 44
order
  partition 22, 39-40
  subtask 24, 61
outnode, operator 44, 70, 76
output

INDEX 219

link 24, 52, 55, 62-63, 67, 70-72, 74-75, 84-86
operator 24, 33, 37, 44, 70, 72
port 24, 52, 55, 62-63, 65-67, 70-72, 74, 82, 84-86, 94, 102
register 24, 28, 33, 52, 65, 67, 70, 72, 74-75, 82, 84, 90, 94, 98
select 22, 39, 44, 53
overhead rule 32
overview, high level 21, 39
packed decimal
    instruction 89
    number 96, 99, 101
pad, temporary 75
page address carrier 44
pair
    attribute value 10, 25
    functional 98
    register 64, 98
parallel
    design 6, 18, 22, 24, 39-40, 48, 69
    implementation 44
    operator 6, 24, 42, 69, 81, 90
parameter
    fold 35
    formal 44, 63, 66, 84
    register 6, 63, 66, 78, 84
    user 50
    user settable 78
parity
    error 98
    generator 96, 98
partition
    estimator 27, 41, 73, 75-77, 80, 87
    experiment 77, 80

high level 39, 41, 58, 80, 87, 90, 101
information 24, 27, 41, 58, 77, 80
order 22, 39-40
path, data 2-7, 18, 22, 27-28, 40, 59, 62, 66, 70-72, 77, 80, 86, 90, 93, 99, 101, 104
pattern
    interpreter 11, 25-26
    shift 101
pending context 34, 55
performance
    designer 77, 80, 82
    rule 61, 105
phase
    clock 18, 22, 40, 66, 93
    clock, two 93
    transition 93
physical
    module 77
    register 42, 44
pipeline, example 3
plan, floor 15, 58, 77, 80, 82, 87, 101, 104
point, floating 89, 94, 99
port
    address 24, 65, 72, 94-95, 102
    bus 55, 84-86, 94-96
    cost 81-82
    destination 25, 62-63
    input 24, 52, 55, 62-63, 65, 67, 70-72, 74, 82-85, 94, 96, 102
    management 58-59, 62-63
    module interface 52
    output 24, 52, 55, 62-63, 65-67, 70-72, 74, 82, 84-86, 94, 102

reference 52, 63, 81-82
route 84
unreferenced 84-85
power 6, 18, 23, 40, 48, 71, 85, 96, 99-100
price
  experiment 82
  hardware 80, 87
primary, memory 94
problem
  complexity, design 69
  constraint 2, 6-7, 23, 96
  division 41, 58
  domain 5, 9-10, 12
  example 9, 16
  solving 5-7, 9, 12, 15, 25, 55
  state 2, 4, 9, 15, 26, 55
procedure
  call 32-33, 38-39, 44
  elicitation 17, 27
process
  design 1, 3, 16, 18, 22-23, 40, 84
  learning 40
  rule selection 26
  selection 11
  synthesis 1, 3
processor
  design 3, 5, 50, 89, 101, 104
  language 3
production system 3, 5, 25
program
  counter 44, 98
  layout 69
  synthesis 3, 5-7, 44
programmable controller 21
programming language 3, 6-7, 10, 15, 25, 42, 77
prototype system 7, 16, 18, 23, 39-40, 77

proximity
  estimator 75-76
  operator 33, 78-80
  register 38, 48, 75, 78-80
  total 78-79, 82
  value 35, 38, 80
purpose register, general 102
R1 12, 14
reasoning
  antecedent 11, 26
  consequent 11
  mechanism 12, 26, 87
reference
  port 52, 63, 81-82
  register 52, 78-79
register
  address 24, 72, 93-94, 96, 98-99
  allocate 7, 22, 24, 33, 40, 58, 63-68, 70, 73, 75, 80, 82-83
  allocation, temporary 70
  architectural 24, 52, 70, 86, 90, 93-94, 98, 102, 104
  array 65, 79, 90, 93-94, 99, 102
  attribute 33, 67, 71-72, 75
  bind 4, 7, 24, 73, 75
  buffer 94, 96, 98, 101
  code 94, 98
  control 2-5, 18, 22, 24, 38, 40, 66-68, 70-72, 74, 78, 86, 94, 96, 104
  data 2-7, 18, 22, 28-30, 40, 66, 72, 86, 98-99, 101, 104
  declared 29-30, 63, 65, 90
  design 6, 21-22, 24, 27, 29, 33, 40, 52, 63, 66-67, 72-75, 79, 82, 84, 86, 90, 99, 103-104
  feeding 33, 74, 84, 93, 96

INDEX 221

general purpose 102
group 96, 98, 102
hardware 6-7, 24, 66, 68, 74-75, 96, 104
input 24, 28, 33, 48, 52, 65, 67, 70, 72, 74-75, 82, 84, 93-94, 96, 98
instruction 45, 67, 93-94, 96, 99
module 4, 24, 48, 52, 58, 64-68, 70-72, 75, 80, 82-84, 86, 104
module, temporary 70, 75
non-architectural 24, 79
operand 98
output 24, 28, 33, 52, 65, 67, 70, 72, 74-75, 82, 84, 90, 94, 98
pair 64, 98
parameter 6, 63, 66, 78, 84
physical 42, 44
proximity 38, 48, 75, 78-80
reference 52, 78-79
remove 24, 66, 70, 72-75, 83-84
stable 24, 38-39, 58, 66, 70, 72-74, 82, 84
status 94, 96, 98-99
storage 24, 28-29, 63, 94, 98
technology-independent 2-4, 41
temporary 27-28, 42, 44, 58, 67-68, 70, 73-75, 93-94, 96, 99, 102
translation 93-94
type 27, 75, 84
unneeded 83
usage 67
working 40, 96, 99
remove

module 61-62, 66, 72, 75-76, 83-84
register 24, 66, 70, 72-75, 83-84
rule 32, 35-36, 38, 61-62, 69-70, 73, 76, 83-85
working memory 32, 36, 69, 73
representation
  abstract 3
  data 6-7, 24, 31, 57
  internal 24, 87
  knowledge 7, 24, 31, 58
  VT 44
  working memory 24, 46-47, 50, 54-55, 87
request
  link 55, 62
  stop 98
rete network 87
rm goal, module 62, 75
route
  connection 41
  port 84
rule
  active 34-38, 85
  allocation 36, 59, 61, 68, 70-71, 82
  antecedent 11, 14, 26, 32, 38-39, 87, 105
  base 33, 105
  bus 29, 33, 83, 85
  change 27, 29, 35, 37, 87, 102, 105
  cleanup 32, 58, 73, 76
  create module 61
  demon 37
  design 4, 26-27, 29, 31-38, 40, 58-59, 61-63, 66, 69, 82, 84, 90, 101, 105

domain-independent 36-39
firing 31, 64, 82
interpreter 10-12, 14-15, 25-26
link 33, 55, 59, 61-63, 70-71,
    84-85
management 59, 61-63
match 11, 26, 33-34, 47
memory 10-12, 14-15, 25-26,
    31, 35-37, 40, 61, 69, 87,
    90, 102-103
modification 33, 37, 90
move module 61
multiple 26, 29
overhead 32
performance 61, 105
remove 32, 35-36, 38, 61-62,
    69-70, 73, 76, 83-85
selection process 26
shrink, design 23, 40
situation action 16
stable 38, 70
total 35, 59, 61, 71, 76
translated 31
type 28, 31-33, 58, 61, 70, 85,
    90
usage 38, 58
writing 33, 61, 87
run time 33, 87, 105
scheme, cache 101
search space 9, 15
select
    block 39, 45, 66
    component 14-15
    goal 61
    operation 53
    output 22, 39, 44, 53
selection
    criteria 31
    process 11

process, rule 26
style 23, 40
sequence, control 2, 34, 52
serial
    design 18, 22, 39-40, 76
    unit 22, 39
service
    function 58-59, 61, 64, 71
    management 63
set
    difference 81
    membership 77, 87
    operation 32, 38, 44
settable parameter, user 78
share hardware 28, 77-79, 81, 87
shift
    operation 24, 42, 71, 98, 100
    pattern 101
show, type 27-28, 62
shrink, design rule 23, 40
silicon chip 32
similarity measure 78
simulate list 32, 38
simulator, implementation 7
single function logic 93
situation
    action rule 16
    error 16
size violation 22, 40
solving, problem 5-7, 9, 12, 15, 25,
    55
space
    design 15, 18, 26
    search 9, 15
special microcode 96, 98
specification
    control 52-53, 63, 67, 95
    input 37, 52-53, 63
    language 7, 52

INDEX 223

model 52
structure 52
specified operator 27, 42, 44, 69
speed violation 22, 40
stable
   calculation 38
   register 24, 38-39, 58, 66, 70, 72-74, 82, 84
   rule 38, 70
state
   initial 26, 83
   match 26, 34
   problem 2, 4, 9, 15, 26, 55
status
   code 34, 94
   context 34, 55
   register 94, 96, 98-99
step
   clock 69
   control 24, 32, 34-35, 38, 44, 47-48, 52, 55, 63, 67, 69, 72, 74, 76, 86, 104
   delay 35, 47-48, 69
   map 44
   micro control 48, 55
stop request 98
storage
   element 6, 24, 63
   global 63
   key 94
   management 98
   register 24, 28-29, 63, 94, 98
   system 93, 98
   wrap 98
store
   address, local 98
   local 41, 99, 102
structure
   bus 22, 83

control 22, 32, 40-41, 52, 69
specification 52
working memory 55, 69, 87
study
   interview, case 16, 18, 20, 39
   method, case 18-19
style
   bus 3, 22, 90, 93, 101
   design 3-4, 6, 18, 23, 40, 90, 104
   implementation 77-78, 80
   selection 23, 40
subgoal 10-11, 61
subtask 3, 24, 26, 34, 58-59, 61, 66, 103
   allocation 58-59, 61, 103
   functional 58
   order 24, 61
summary, development 39
supplied constraint 2, 64
switch, context 72
symbolic microcode 24, 95
synchronization, clock 50
synchronous design 74, 87, 104
synthesis
   algorithm 7
   design 1-7, 18, 24, 44, 75-76, 95, 102-104
   environment 7, 104
   implementation 2, 7, 18
   level 3-4
   operation 24, 102, 104
   process 1, 3
   program 3, 5-7, 44
   task 2-4, 7, 24, 95
   technique 4, 102-103
   tool 1, 3-4, 7
   use 2-3, 7, 24
system

description 1-2, 42, 52
design 1-7, 15, 18, 21, 23, 26-27, 87, 89, 102-105
development 1-3, 7, 9, 15-16, 39-40, 105
diagnostic 12
domain 4, 9-10, 12
knowledge acquisition 105
knowledge based expert 2, 4, 7, 9-10, 12, 15-16, 23, 31, 40, 103-105
memory 10-12, 21, 87
production 3, 5, 25
prototype 7, 16, 18, 23, 39-40, 77
storage 93, 98
System/370 5, 7, 31, 58, 89-90, 93, 96, 98-102, 104
 architecture 93, 98-99, 101
 design 5, 7, 90, 100, 104
 experiment 31, 102
task
 allocation 17-18, 39, 61, 66, 86, 103-104
 control 34, 41, 61, 66, 86, 103-104
 knowledge 5, 7, 11, 15, 17, 26, 32, 39, 103
 synthesis 2-4, 7, 24, 95
teaching environment 105
team design 21, 90, 99
technique
 design 4-5, 18, 21, 26, 35, 69, 102-103
 interview 5, 18, 26
 match 102-103
 synthesis 4, 102-103
 testing 4, 70
technology
 change 22, 40, 87
 consideration 7, 21, 39, 47
 constraint 7, 18, 21-22, 39-40, 47-48, 64, 100
 database 47-48, 58, 64, 66
 design 1, 3, 7, 21-22, 39-40, 48, 99-100
 fabrication 1
technology-independent
 database 4, 32, 35
 description 2, 24, 32, 41, 52
 hardware-network level 5, 77, 80
 module 4, 24, 80
 register 2-4, 41
technology-sensitive
 constant 87
 database 66, 69
 factor 77, 79-80
 information 47-48, 77, 80
technology-specific hardware 102
temporary
 carrier 42, 44
 module 67, 70, 75-76
 pad 75
 register 27-28, 42, 44, 58, 67-68, 70, 73-75, 93-94, 96, 99, 102
 register allocation 70
 register module 70, 75
 value 34, 37, 44, 67, 70
testability 6-7, 67, 76, 87, 104
testing technique 4, 70
time
 design 1-2, 18, 27, 35, 88, 90, 102-104
 execution 87, 103
 information 2, 20, 86
 modification 33

run 33, 87, 105
tool
    design 1, 3-4, 7, 15, 23, 61, 103
    synthesis 1, 3-4, 7
total
    proximity 78-79, 82
    rule 35, 59, 61, 71, 76
trace, value 3, 33-37, 47, 58, 66
transform operation 32, 71
transition, phase 93
translated rule 31
translation
    hardware 101
    name 47
    register 93-94
trap handler 98
tree, multiplexer 83-84
trim
    input 70, 72
    operator 37, 55, 58, 69-70, 72
two phase clock 93
type
    arithmetic 35
    attribute 25, 44, 66, 75-76
    knowledge 5, 15, 31-32, 58, 61
    module 25, 28, 33, 52, 61, 66, 76, 80, 84
    operator 25, 27-28, 37, 44, 70, 77, 79
    register 27, 75, 84
    rule 28, 31-33, 58, 61, 70, 85, 90
    show 27-28, 62
unbind module 33
unit, serial 22, 39
University, Carnegie-Mellon 1, 89
unneeded register 83
unpack instruction 96

unreferenced
    goal 84-86
    module 83-85
    port 84-85
unused
    constant 83
    module 82
usable design 4, 15, 104
usage
    bus 85
    hardware 6, 90
    register 67
    rule 38, 58
use, synthesis 2-3, 7, 24
user
    interaction 104
    parameter 50
    settable parameter 78
value
    delay 35, 47, 50, 69
    input 33, 44, 53, 70, 74, 84
    pair, attribute 10, 25
    proximity 35, 38, 80
    temporary 34, 37, 44, 67, 70
    trace 3, 33-37, 47, 58, 66
variable allocation 34-36, 65, 70
violation
    constraint 18, 22, 24, 40
    size 22, 40
    speed 22, 40
virtual address 98, 101
VT
    allocation 58, 63, 67, 72, 82
    description 89
    representation 44
weak method 17, 26
weight
    average 78-79, 81-82
    factor 77, 79-80

wiring
  module 52, 67, 82
  operation 71, 73, 84
word, microcode 95, 98
working
  chip 32, 99-100
  memory 10-12, 15, 21, 25-26,
    31, 35-37, 61, 69, 76, 87, 96
  memory cleanup 32, 73, 76
  memory list 32, 38, 46-47, 50,
    54, 61, 69
  memory, remove 32, 36, 69, 73
  memory representation 24, 46-
    47, 50, 54-55, 87
  memory structure 55, 69, 87
  register 40, 96, 99
world knowledge 31, 40
wrap, storage 98
writing rule 33, 61, 87